PHYSICS SQUARED
100 CONCEPTS YOU SHOULD KNOW

GRAHAM SOUTHORN
AND GILES SPARROW

APPLE

www.apple-press.com

ISBN 978-1-84543-646-9

QUMPHSQ

This book was conceived, designed and produced by
Quantum Books Limited
6 Blundell Street
London N7 9BH
United Kingdom

Publisher: Kerry Enzor
Editorial and Design: Pikaia Imaging
Editor: Anna Southgate
Design: Dave Jones
Illustration: Tim Brown
Production Manager: Zarni Win

Printed in China by Toppan Leefung Printing Limited

9 8 7 6 5 4 3 2 1

Cover image: (c) itVega/Shutterstock

Contents

Introduction

The Nobel prize-winning physicist Richard Feynman was
frequently asked to explain the science that won him the
prestigious award. 'If I could explain it to the average person,
it wouldn't have been worth the Nobel prize,' he once joked
in response. At the highest level, physics can indeed be hard
to understand. But at its heart, physics is actually about
the search for the simplest possible way of describing the
workings of Nature. Such simple descriptions are often
written in mathematical terms that can look daunting to the
uninitiated, but not all physics equations are complicated.
Perhaps the simplest of all, Albert Einstein's famous $E = mc^2$,
is also one of the most powerful.

Nonetheless, we believe it is possible to grasp the essential
concepts of physics without looking at equations, and
that's what we've attempted in this book. Here, each of
ten chapters, is further divided into ten bite-sized topics.
Collectively, they attempt to summarize the key areas of
physics – from the tiniest particles inside atoms to the
gravitational forces at work across the entire universe.

The useful fact that mathematics can describe the world
so accurately lies at the heart of physics, and it's led to

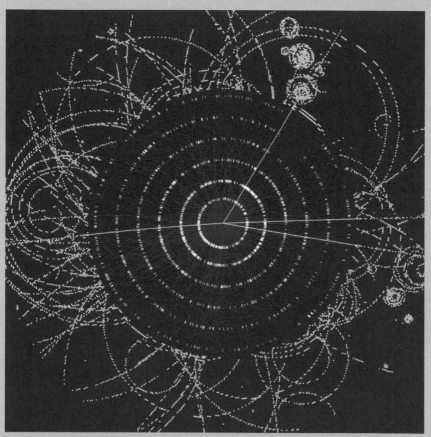

In 1964 Peter Higgs and other physicists predicted the existence of a mechanism through which other particles got their mass. Some 50 years later, the Higgs boson (shown here through computer-modelled tracks) was detected at the Large Hadron Collider, reinforcing the theory.

an ongoing game of tag between theoreticians, who come up with ideas, and physicists who perform experiments. Theories, underpinned by mathematics, not only describe known phenomena, but also predict new ones. Today, thanks to more than three centuries of theory and experiment, physics has laws that explain the behaviour of particles inside atoms (quantum theory) and black holes (general relativity). Newton's laws describe our everyday experience of motion, while the laws of thermodynamics describe heat. Electricity, magnetism and light, meanwhile, are all part of one theory, developed by James Clerk Maxwell, which depicts all three phenomena as waves. We now know more about what everything is made of, thanks to enormous particle accelerators. These have revealed families of subatomic particles, as well as the rules governing how they interact.

But what, in the end, is physics for? Why does it matter if a bunch of mathematical equations tell you how a ray of light will refract through glass, or how heat is transferred through a metal? The answer is simple: understanding how the world works gives us the means of exploiting it, and even the most complicated areas of blue-sky research can ultimately yield surprising practical advances. Without the advances made by generations of physicists, our lives would be immeasurably poorer.

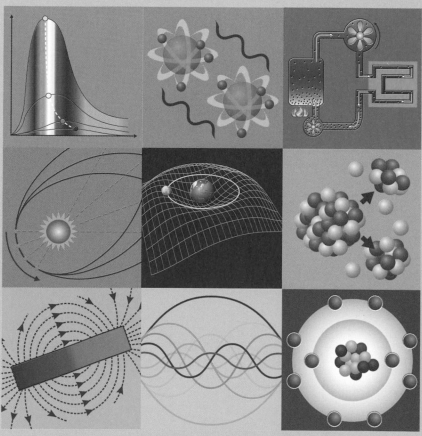

Among the subjects covered in this book are the following (as illustrated from top left to bottom right): the black body curve, entanglement, heat engines, gravity, spacetime, fission energy, magnetism, harmonics and atomic structure.

MECHANICS
AND MOVEMENT

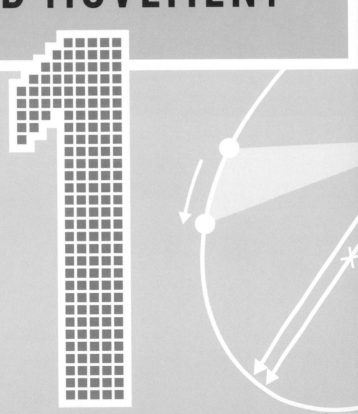

Of all the physical forces that control the universe and the world around us, those of mechanics are by far the most obvious. Each of us has an innate knowledge of mechanics – learned since birth – that comes into play every time we judge the distance of approaching traffic or play a sport, be it anything from pool to Frisbee.

Mechanics is the oldest branch of physics. It is concerned with the way in which objects move and interact with their environments when subjected to external forces. In practice, this definition covers a surprising range of physical systems and situations. Obvious scenarios include the path of a pool ball once in motion or the flight path of an accelerating rocket. There are other cases, however, such as the bulk behaviour of gases, in which the mechanics are hidden out of sight.

Continues overleaf

The laws of mechanics underpin many other branches of physics. This chapter takes a fresh look at some of the basic concepts before drawing out the rules that unite them. It also considers key applications of mechanics in fields such as engineering.

The fundamentals of mechanics may be long-established, but this should not undermine their significance. It is clear that some of the systems that follow these apparently simple rules are complex and chaotic. They are also impossible to predict without an infinite amount of information, as many weather forecasters can attest. What's more, we now know that these classical rules are not the whole story. In extreme situations they break down, and the strange rules of quantum mechanics and relativity take their place (see Chapters 8 and 10).

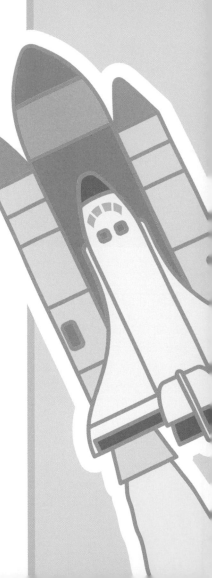

Contents

1.1 Speed, velocity and acceleration

It is possible to calculate an object's speed, velocity and acceleration using simple mathematics.

Whether thrown, dropped or powered in some other way, many objects in the universe – and in our immediate environment – move around. We describe such motion using three principal terms.

- Speed is simply the change in an object's position over time. A car's speed is usually given in km/miles per hour. In order to work out a car's speed, you simply divide the distance travelled by the time it took to get there.

- Velocity is speed in a particular direction. This makes velocity a **vector quantity** – that is, it has both magnitude and direction. This comes in handy for working out **relative velocity**. For example, if a car travelling at 48 km/h (30 mph) hits another vehicle head-on, which is travelling straight towards it at 32 km/h (20 mph), the relative velocity is 48 + 32 = 80 km/h (50 mph). This is why head-on collisions are so deadly.

- Acceleration is the rate of change in velocity. It tells you how quickly velocity is increasing every second (or decreasing if an object is decelerating). It's usually given in units such as metres per second per second, or metres per second squared.

Knowing how to calculate speed and acceleration allows us to make valuable comparisons between, say, different vehicles or human athletes competing for medals.

A rocket has to reach an escape velocity of 11.2 km (7 miles) per second in order to overcome the earthward attraction caused by gravity.

Mapping velocity

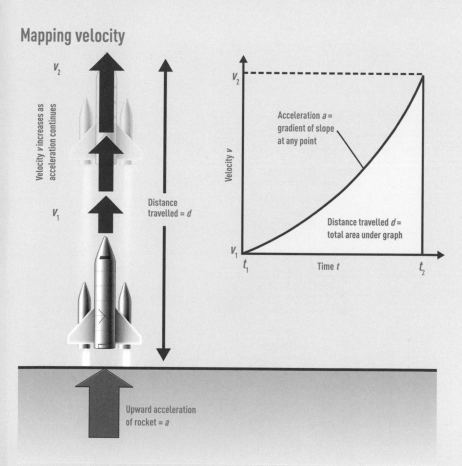

v_2

Velocity v increases as acceleration continues

v_1

Distance travelled = d

Velocity v

v_2

Acceleration a = gradient of slope at any point

Distance travelled d = total area under graph

v_1

t_1 Time t t_2

Upward acceleration of rocket = a

Motion is often mapped using graphs. Here, the velocity of a rocket launching into space is mapped on a simple graph, which also yields its rate of acceleration and total distance travelled (shaded yellow).

1.2 Inertia, mass and weight

Why is it that an astronaut standing on the Moon weighs less than that same astronaut standing on Earth?

When in motion, an object will carry on moving at the same velocity and in the same direction unless a different force acts upon it. This concept is known as **inertia** and it's one of Isaac Newton's laws of motion (see Topic 1.4).

The same laws say that an object that's not moving will continue to remain stationary. This is where another property comes into play: mass. Mass is how much matter an object contains, measured in kilograms or pounds.

The more mass something has, the greater the force needed to change its velocity. Think of something massive, like a large truck. It takes a much greater force to get a heavy truck to move, or to slow down, than it does a car.

In everyday life, we generally say weight when we're referring to mass. But in physics, weight is a different concept altogether. Measured in newtons (N), weight is the force that acts on an object owing to gravity.

An object has the same mass wherever it is, but its weight varies from place to place depending on gravity. For example, gravity is lower on the Moon than it is on Earth, which means weights are lower, too. It is for this reason that the weight of an astronaut standing on the Moon is only one-sixth of that same person's weight on Earth.

Isaac Newton called inertia the innate force of matter.

Mass versus weight

On Earth, a 907-kg (2,000-lb) mass has a weight of 8,900 N, because the force of gravity on Earth is 9.81 N/kg (4.45 N/lb)

907 kg

907 kg

On the Moon, that same mass has a weight of 1,460 N, because the force of gravity on the Moon is lower, at 1.62 N/kg (0.73 N/lb)

An object's mass remains the same wherever it is – the gravitational field has no influence. However, that same object's weight – the force gravity exerts – can vary dramatically, as illustrated above.

1.3 Momentum and collisions

Different forces come into play when one moving object hits another, or even something that is not moving at all.

A moving object has a property that depends on both its velocity and mass. This property is **momentum**. In Topic 1.2 we discussed a large truck being hard to stop. Momentum is a measure of just how hard, and is calculated by multiplying an object's mass by its velocity. The result is given in units such as kilogram metres per second (abbreviated to kg m/s).

Like velocity, momentum is a vector quantity with a direction. Importantly, momentum is the same before and after a collision. In other words, it is **conserved**. This is an important factor when working out what happens to an object in motion.

Consider a game of pool. When the cue ball fires into the pack it may stop completely itself, while other balls in the pack ricochet off, albeit at lower speeds. The momentum is conserved – in this case – in a straight line, and is called **linear momentum**. The similar property of a body spinning around an axis, or in orbit, and maintaining momentum is called **angular momentum** (see Topic 1.5).

Conservation of momentum even comes into play when you throw a ball or stone. Since momentum is zero to start with, you gain backward momentum once you've thrown the stone. In practice, you won't notice this owing to the effect of friction caused by the ground – unless you happen to be standing on ice, that is.

In physical equations, momentum is usually denoted by the letter *p*, for *petere*, the Latin word for 'push'.

This controlled car crash reveals the forces generated when
a less massive, fast-moving object with high momentum collides
with a more massive, static object that has no momentum at all.

1.4 Newton's laws of motion

From trains and cars to spinning plates and falling apples, here are three simple laws that explain the movement of things in the world around us.

In 1687, the English scientist Isaac Newton devised three laws to explain the behaviour of an object when in motion. Today, his laws of motion form the bedrock of what is referred to as classical physics. The laws comprise three simple statements:

■ Any object will remain in a state of uniform motion, or at rest, unless acted upon by an external force.

■ The acceleration of an object is directly proportional to the forces acting upon it, but inversely proportional to its mass. Acceleration is easy to calculate using a simple equation – you just divide force (F) by mass (m): $a = F/m$.

■ Every action has an equal and opposite reaction. In other words, the force generated by one body on another is matched by an equal and opposite force exerted by the second body on the first.

These three laws describe the movement of everything familiar in the world around us. However, centuries after Newton, other scientists discovered that these laws tend not to apply in extreme circumstances, and this led to the exploration of quantum mechanics and relativity (see Chapters 8 and 10).

Despite his reputation as one of history's greatest scientists, Newton was also fascinated by the mystical practice of alchemy.

Newton's laws in action

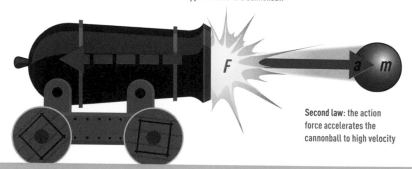

First law: an explosion applies force to a cannonball

F

a m

Second law: the action force accelerates the cannonball to high velocity

Third law: a reaction force pushes the cannon backwards

According to Newton's laws, the system of cannon and cannonball remains at rest until the explosive force is applied (first law). The action force accelerates the cannonball by a given amount ($a = F/m$; second law), while a reaction force causes the cannon to recoil (third law).

1.5 Orbits and angular momentum

There is a perfect balance between the forward motion of the planets and gravitational pull from the Sun.

The physical properties of motion do not apply to objects on Earth alone, but to those everywhere in the universe. They also apply to the planets orbiting the Sun, therefore, and to the satellites orbiting the planets.

An **orbit** is the path taken by one object around another. In 1609 Johannes Kepler devised three laws that describe the orbit of planets around the Sun, the first of which is the focus here. It says that all planetary orbits are ellipses. This had not been obvious previously, because the ellipses are almost circles, but Kepler relied on careful measurements made by astronomer Tycho Brahe to establish this law.

The force needed to keep planets orbiting the Sun is gravity. This **attractive force** between objects is determined by their masses and the distance between them. Without gravity, a planet would simply continue in a forward motion. However, because the Sun's mass is so large, its gravity introduces just enough pull for the planets to maintain their orbits.

Objects moving on curved paths also have angular momentum (see Topic 1.3). This property is linked to an object's mass, its angular velocity and distance from the centre of rotation. It explains why figure skaters spin faster when they tuck in their arms: the distance travelled is reduced, while angular velocity increases in order to keep angular momentum the same.

Kepler's laws were the final piece of evidence that clinched the case for a Sun-centred Solar System.

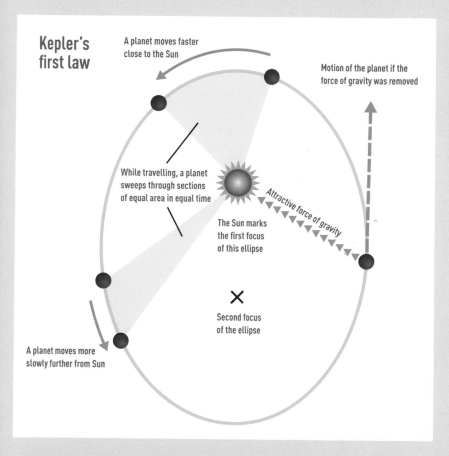

Kepler's first law

A planet moves faster close to the Sun

Motion of the planet if the force of gravity was removed

While travelling, a planet sweeps through sections of equal area in equal time

Attractive force of gravity

The Sun marks the first focus of this ellipse

✕
Second focus of the ellipse

A planet moves more slowly further from Sun

Kepler's laws describe how planets move around the Sun on elliptical paths. It was Isaac Newton who identified gravity as the universal force responsible for these paths.

1.6 Work, energy and power

When a force moves an object, there is more going on than the laws of motion describe.

In physics, the three concepts of work, energy and power quantify the effort required to make something move. They also explain how energy is transferred between objects or systems and at what rate.

■ When a force moves an object, the work done is that force multiplied by distance. The work done by lifting a weight of 20 N by 2 m (80 in), for example, is 40 newton metres. This is equal to 40 joules – a unit not just of work, but of energy, too.

■ Energy in physics is an object's or system's capacity to do work. For example, consider how it saps your energy to lift objects or climb stairs. Energy cannot be created or destroyed, but it can be transferred from one form to another.

Types of energy we encounter all the time include **kinetic energy**, which is due to an object's motion, and **potential energy**, which is due to position. The potential energy of a skier at the top of a slope is transferred to kinetic energy with their downward motion.

■ Power is the rate at which energy is expended, and is measured in watts (joules per second). A way of quantifying power was crucial for demonstrating the benefits of steam engines, because it allowed their output to be compared to that of horses.

Watts and joules are also used for measuring electrical energy and power.

Energy transfer

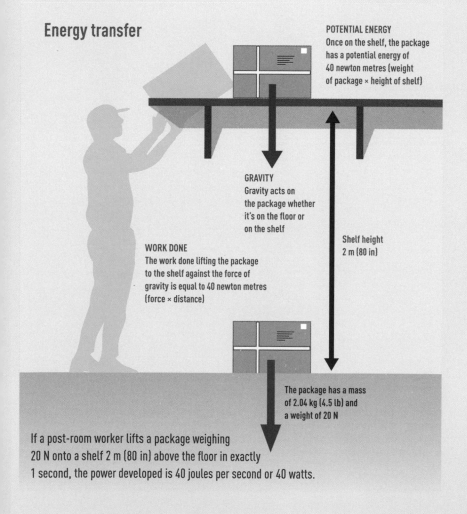

POTENTIAL ENERGY
Once on the shelf, the package has a potential energy of 40 newton metres (weight of package × height of shelf)

GRAVITY
Gravity acts on the package whether it's on the floor or on the shelf

Shelf height
2 m (80 in)

WORK DONE
The work done lifting the package to the shelf against the force of gravity is equal to 40 newton metres (force × distance)

The package has a mass of 2.04 kg (4.5 lb) and a weight of 20 N

If a post-room worker lifts a package weighing 20 N onto a shelf 2 m (80 in) above the floor in exactly 1 second, the power developed is 40 joules per second or 40 watts.

1.7 Simple machines

A variety of familiar devices make tasks easier by reducing the effort required to do work.

When we think of machines we imagine steam engines, cars and aeroplanes. In physics, a machine is simply a device that provides some kind of **mechanical advantage**, in order to reduce the amount of force needed to do a certain amount of work.

The equation, work = force × distance moved tells us that, in order to achieve the same amount of work, we can apply half the force over twice the distance – or double the force over half the distance. This concept is central to the function of many simple machines.

Take, for example, a lever. If you place a heavy weight on one end of the beam, the easiest way to raise it up in the air is to place the pivot close to that end and apply force to the opposite end. The distance from the force to the pivot means less force is required than if the pivot was centred.

Other kinds of simple machines include inclined planes and wheels. It's much easier to push a heavy object up an inclined plane, like a ramp, than it is to lift it. You'll do the same amount of work overall, but using less force, because the object travels further. A wheel and axle allows you to exert a relatively small force through a large distance on the rim, in order to lift a heavy weight attached to the axle.

The rock axe – the earliest tool – was invented by human ancestors at least two million years ago.

Levers, inclines, wheels and axles

$$MA = \frac{Length}{Rise}$$

LEVER

$$MA = L_2/L_1$$

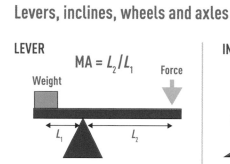

When using a lever, the mechanical advantage is calculated by dividing the distance from the force to the pivot, by the distance from the pivot to the weight – in this case the force is doubled

INCLINE

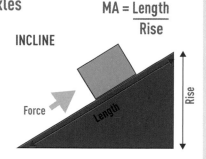

When using an incline, the mechanical advantage can be calculated by dividing the distance travelled by the distance the weight is raised

WHEEL AND AXLE

$$MA = R/r$$

When using a wheel and axle, the mechanical advantage is calculated by dividing the radius of the wheel by that of the axle

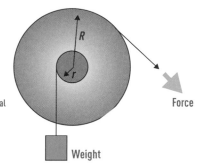

The mechanical advantage (MA) of a simple machine is the factor by which a machine multiplies the force applied to it. These diagrams show how to calculate it for different machines.

1.8 Deformation, strain and stress

**When forces act
on fixed objects,
the result is often
a change in shape.**

So far we have seen how objects move around and what
happens when their energy is transferred. But not all
objects are like pool balls, which retain their shape and
size when they collide. In fact, most objects become
deformed in some way, as their atoms and molecules
become stretched and rearranged.

There are different types of deformation. In **elastic
deformation**, an object recovers its original shape once the
force is removed. Think of a spring that bounces back after
being compressed. Hooke's law (Robert Hooke, 1660),
states that the extent of the deformation is proportional
to the force applied. If you hang a weight on the end of a
spring and take it off again, the spring will snap back to its
original length – but only if the weight isn't too heavy.

A heavier weight will stretch the spring so far that it will
no longer snap back. This is **plastic deformation** – a
permanent alteration that occurs at the elastic limit of the
material. Apply yet more force, and the material will break
or crack, known as a fracture.

**Brittle materials
progress from elastic
deformation to fracture
with very little plastic
deformation in between.**

When deformation occurs, **strain** is the change in length
of a material, compared to its original length. **Stress** is the
force per unit area required to cause that change. These
concepts are essential to engineers, and allow them to
quantify how different materials deform.

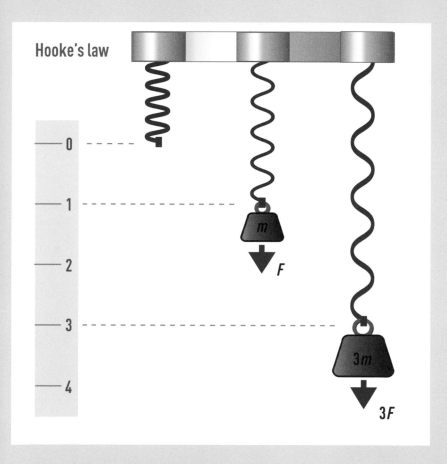

Hooke's law

- 0
- 1
- 2
- 3
- 4

m

F

$3m$

$3F$

According to Hooke's law the extent of deformation – in this case, the increasing length of a spring – is proportional to the amount of force (F) applied, as shown here, by hanging mass (m) from the bottom.

1.9 Gas laws

A gas is made up of numerous atoms or molecules. Its behaviour is best understood by modelling the mechanics of these particles.

The laws of physics do not only apply to solid objects. There are also laws governing the behaviour of gases. They demonstrate how the various properties of a gas relate to one another.

Unlike solids, the atoms and molecules in a gas are not bound together, and so they expand to fill the volume of a container. From the 17th century onwards, scientists discovered three laws relating to the pressure, volume and temperature of a gas (measured on the Kelvin scale; see Topic 3.5), each of which can be expressed using a simple equation.

- **Boyle's law** states that pressure is inversely proportional to volume in a gas at constant temperature ($p \propto 1/V$).

- **Charles's law** states that volume is proportional to a gas's temperature at constant pressure ($V \propto T$).

- **Gay-Lussac's law** states that pressure is proportional to absolute temperature at constant volume ($p \propto T$).

The three laws combine to produce one single equation, $pV \propto T$ – the product of a gas's pressure and volume is proportional to its temperature. This applies to all simplified, ideal gases. The consequences of these laws are seen everywhere, from hot-air balloons to pistons and in the internal combustion engine.

Avogadro's law states that the volume occupied by a gas in standard conditions is proportional to the number of molecules it contains.

Gas laws in action

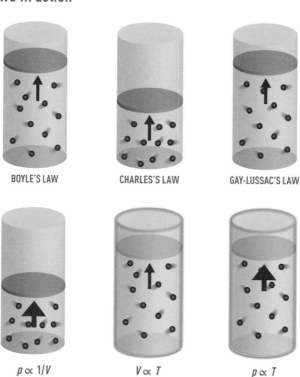

BOYLE'S LAW   CHARLES'S LAW   GAY-LUSSAC'S LAW

$p \propto 1/V$ $V \propto T$ $p \propto T$

The three relationships described by Boyle's law, Charles's law and Gay-Lussac's law. The red arrows indicate the amount of pressure exerted, and increased temperature is indicated by a warmer tone.

1.10 Fluids

No matter how thick it is, or how slowly it travels, a fluid is any substance that changes shape at a steady rate under the influence of an appropriate force.

When we think of fluids, we generally imagine soft drinks, rivers and oil. In physics, however, a fluid is anything that can flow. This includes liquids, but also gases and even some solids.

The study of fluids goes back to the ancient Greek scientist Archimedes and his famous Eureka moment. Getting into his bath, he discovered buoyancy – the idea that the upward force on a body immersed in a fluid is equal to the weight of the fluid being displaced. This concept explains why some objects float while others sink.

Towards the end of the 17th century, Isaac Newton became the first scientist to measure another property of fluids: viscosity, or the extent to which a fluid resists flowing. In a liquid, we tend to think of viscosity as that liquid's thickness – syrup is more viscous than water, for example.

During the first half of the 18th century, the Swiss scientist Daniel Bernoulli discovered that fast-flowing fluids exert less pressure than slow-flowing ones. This so-called Bernoulli effect is the principle behind the design of an aircraft's wings. Their shape exploits a velocity difference in the air passing above and beneath the wings in order to produce a pressure difference, thus lifting the aeroplane.

Formula 1 race-car designers use the Bernoulli effect in reverse, creating downforce that increases a car's grip on the track.

These general principles apply to all fluids and underpin many aspects of everyday life.

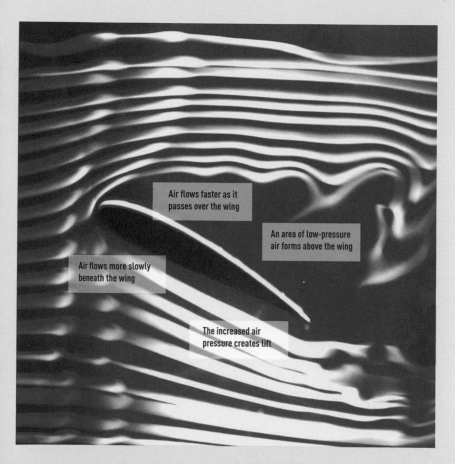

A wind-tunnel shows the Bernoulli effect in action. The profile of an aircraft wing diverts the flow of air, creating a faster-flowing, lower-pressure area above the wing, producing lift and also turbulence.

WAVES

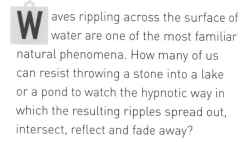

Waves rippling across the surface of water are one of the most familiar natural phenomena. How many of us can resist throwing a stone into a lake or a pond to watch the hypnotic way in which the resulting ripples spread out, intersect, reflect and fade away?

Yet water waves are just the most obvious manifestation of a far more fundamental phenomenon – waves are widespread in nature, often hidden away in forms we cannot see. The sounds that make our eardrums vibrate, for example, are simply caused by a certain type of wave moving through air. Vision relies on another type, which stimulates sensitive cells in our eyes.

Waves are all around us and behave according to a set of rules that are distinct from those affecting particles. A proper understanding of these rules is important to the foundations of physics.

Continues overleaf

Scientifically speaking, a wave is a periodic disturbance that transfers energy from one place to another via one substance or another. Waves usually have certain shared characteristics, such as wavelength across which the disturbance repeats itself.

We often think of wave-carrying media in terms of fluids – usually water or air – but most materials, including solids, will transmit sound waves. Even the earth itself vibrates when its internal rocks shift and grind past one another. However, one special wave – electromagnetism – has a unique way of travelling through space without the need for a transmitting medium (see Chapter 4). In order to understand this wave's unique properties and technological importance, we must first get to grips with the fundamental properties of waves.

Contents

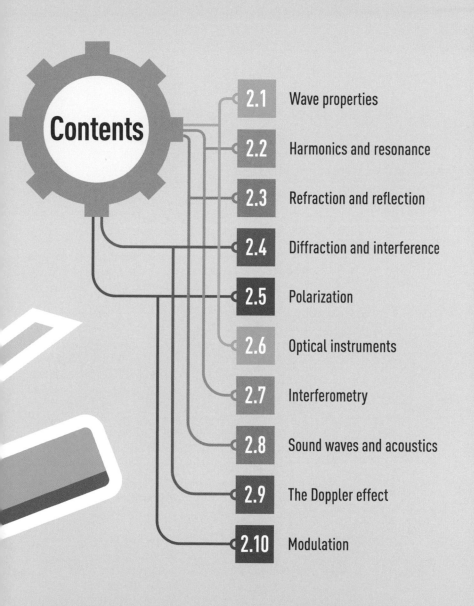

2.1 Wave properties

Waves are defined by their speed, frequency and wavelength, all of which are measured in terms of their cyclical wave fronts.

Waves don't just occur in the sea. Different kinds of waves travel through solids, gases and vacuums as well as liquids. All of them are governed by the same laws of physics. A wave is simply a repeating disturbance that travels through space and carries energy.

■ **Transverse waves** are characterized by a displacement at right angles to the direction in which the wave is moving. Imagine a Mexican wave at a soccer stadium. The wave travels around the stadium but the displacement involves people jumping up and down.

■ **Longitudinal waves** are characterized by a displacement parallel to the direction of the wave. They are caused by compression creating areas of high and low pressure along the wave's path. Sound waves are longitudinal waves.

Both kinds of waves have three common properties. The first is **wavelength**. In transverse waves, this is the distance between one peak and the next. In longitudinal waves it is the distance between points of greatest compression. **Frequency**, measured in cycles per second, or hertz (Hz), is the rate at which waves are produced. **Amplitude** describes the maximum displacement at a peak or trough, or the greatest amount of compression.

The average human ear hears sounds in a range from about 20 to 20,000 Hz.

Usefully, you can calculate the **velocity** of a wave just by knowing its wavelength and frequency.

Characteristics of a wave

TRANSVERSE WAVE

Wavelength = distance between successive peaks or troughs

Frequency = rate of wave production

Amplitude = greatest distance from centre line

LONGITUDINAL WAVE

Amplitude = greatest pressure change from normal

Frequency = rate of wave production

Wavelength = distance between successive areas of greatest compression or rarefaction

One way of understanding the different types of wave is to imagine them applied to a child's spring toy, the Slinky, as shown above. Multiplying a wave's frequency by its wavelength gives its speed of travel through space.

2.2 Harmonics and resonance

A series of waves
whose wavelengths
are neat fractions
of each other makes
sweet music.

Having a good grasp of the physics relating to waves helps us to understand two properties encountered in music: harmonics and resonance.

When a wave travels through a medium that is fixed at both ends – say, a guitar string – the lowest possible frequency is called the **fundamental frequency**. It corresponds to a wavelength (l) that is twice as long as the string. Written as a formula, this is $l = 2d$. The amplitude is zero at either end of the medium and peaks in the middle.

Pluck a guitar string and you'll hear the fundamental frequency. If you hold the string down halfway along its length, at its midpoint, you create two lengths of string. When plucked, each doubles the frequency of the original length. Further multiples of the fundamental frequency form a complementary set of harmonics, which produce pitches that are pleasing to the human ear.

These multiples are **resonant frequencies**. As waves travel back and forth, some are reinforced while others cancel each other out. Resonance is also seen in everyday objects.

This is why your furniture might vibrate when you turn on a vacuum cleaner – the motor generates waves that have the same frequency as the furniture's natural frequency.

In 1940, the Tacoma Narrows Bridge in Washington State, USA, collapsed after winds set off vibrations at its resonant frequencies.

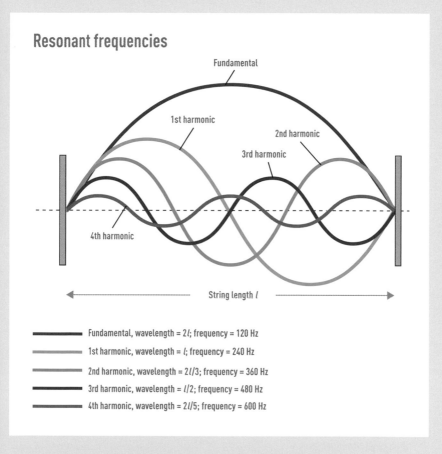

Resonant frequencies

Fundamental

1st harmonic

2nd harmonic

3rd harmonic

4th harmonic

String length *l*

Fundamental, wavelength = 2*l*; frequency = 120 Hz

1st harmonic, wavelength = *l*; frequency = 240 Hz

2nd harmonic, wavelength = 2*l*/3; frequency = 360 Hz

3rd harmonic, wavelength = *l*/2; frequency = 480 Hz

4th harmonic, wavelength = 2*l*/5; frequency = 600 Hz

This schematic shows the fundamental frequency and first few harmonics for a plucked string with a fixed length *l*. Harmonics are most familiar from music but similar phenomena can occur in any wave-carrying medium.

2.3 Refraction and reflection

Waves change direction when they meet a boundary between two media.

Waves don't always travel through the same medium, but sometimes move from one to another. When a wave encounters a boundary between two media, one of two things can happen. **Refraction** is when the wave continues into the new medium with a changed direction. **Reflection** is when the wave bounces back. Both depend on the angle at which the wave hits the boundary, and the nature of the two media.

When light travels into a different medium, the extent to which the light bends is determined by the **refractive index** of both media. The refractive index is the amount by which light slows down in a given medium compared to its speed in a vacuum. A vacuum has a refractive index of 1, for example, and water of around 1.3. One law of physics, called Snell's law, relates the angles at which light enters and leaves a new medium to the refractive indices of both media (see opposite). This bending of light explains why a swimming pool looks shallower than it is.

Most reflecting surfaces actually absorb or refract some of a wave's energy – this why reflections in mirrors look slightly dimmer than in real life.

In 1678, Christiaan Huygens came up with a theory to explain how waves move through space. Huygens' principle holds that every point on the surface of a wave front, is a source of secondary waves called wavelets. When a wave strikes a new medium, one part of the wave front generates wavelets before the rest, and it is this that causes a change in direction.

Angles of refraction

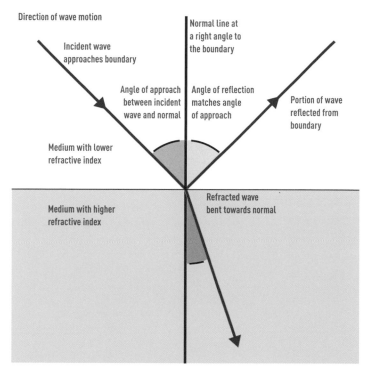

Direction of wave motion

Incident wave
approaches boundary

Normal line at
a right angle to
the boundary

Angle of approach
between incident
wave and normal

Angle of reflection
matches angle
of approach

Portion of wave
reflected from
boundary

Medium with lower
refractive index

Medium with higher
refractive index

Refracted wave
bent towards normal

A refracted wave is always bent towards the normal when it enters
a medium with a higher refractive index. It bends away from the
normal when it enters a medium with a lower index.

2.4 Diffraction and interference

When waves combine with each other, they either increase in intensity or cancel each other out.

In Topic 2.3, we saw that Huygens' principle describes waves by imagining each point on an advancing wave front as a source of wavelets. Wavelets spread out in all directions, but only reinforce each other in the direction of travel. This idea is useful when it comes to understanding two other wave phenomena.

- **Diffraction** occurs when a wave's path is blocked by a barrier, such as a breakwater in the case of ocean waves. Even if the waves are perfectly parallel on one side of the barrier, they tend to spread out on the other side. You can think of this effect in terms of new wavelets spreading out from the edges of a wave once it has passed through a gap in the breakwater. Similarly, a light that shines through a small hole in a piece of card is diffused, rather than forming a precise beam.

- **Interference** occurs when one wave runs into another. This means that when two wave peaks overlap they reinforce each other (**constructive interference**). When a peak encounters a trough, however, the wave is cancelled out (**destructive interference**). This was clearly demonstrated in a classic experiment in 1801 by Thomas Young (see opposite). He showed that light passing through two narrow slits produced alternating bands of light and dark on a screen. The paired beams of light emerging from the slits were interfering with each other.

Noise cancellation systems use destructive interference to cancel out unwanted sound waves.

The double-slit experiment

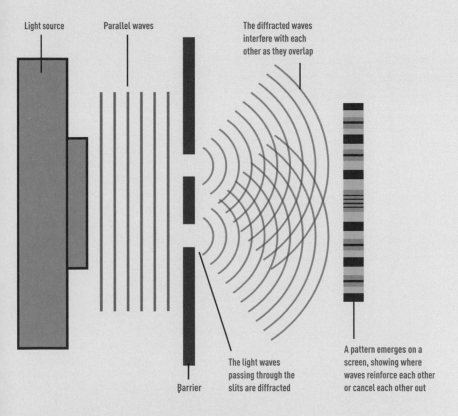

Light source

Parallel waves

The diffracted waves interfere with each other as they overlap

The light waves passing through the slits are diffracted

Barrier

A pattern emerges on a screen, showing where waves reinforce each other or cancel each other out

Young's double-slit experiment proved that light diffracts and interferes in the same way as water waves, creating a pattern of constructive and destructive interference.

2.5 Polarization

It is possible to filter light waves in order to block some of the light passing through.

The waves in water are transverse waves. Their oscillations are oriented in a fixed direction (up and down), which is perpendicular to the direction of travel (see Topic 2.1). Electromagnetic waves carry an oscillating electric field that doesn't always point the same way. For example, in natural sunlight, the electric field oscillates in multiple different planes in different rays of light.

This property is exploited in the technology behind sunglasses and LCD TV screens, because it is possible to **polarize** light so that it vibrates in just one plane. This is achieved by passing the light through a polarizing filter (see opposite). Polarizing filters consist of a microscopic grille made of crystals. The crystals are aligned in such a way that the grille only allows light through if it's travelling in a specific direction.

Most light rays are polarized at random; the polarizing filters used in sunglasses reduce the amplitude to half its original strength, on average. If two polarizing filters are aligned at 90 degrees to one another, however, almost all incoming light is blocked.

In LCD TVs and digital camera screens, the polarization of sheets of liquid crystal can be changed by applying a voltage. This lets through, or blocks, illumination from a backlight to create light and dark pixels in the desired locations to form an image.

Some natural crystals send light in different directions depending on its polarization.

Polarized 3-D

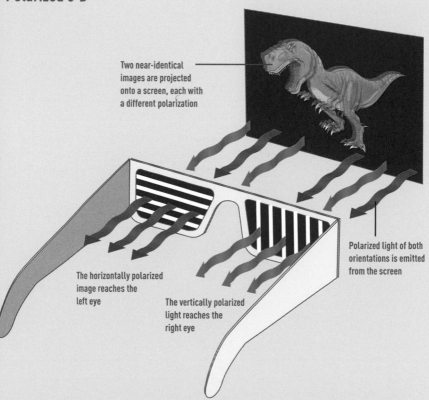

Two near-identical images are projected onto a screen, each with a different polarization

Polarized light of both orientations is emitted from the screen

The horizontally polarized image reaches the left eye

The vertically polarized light reaches the right eye

A popular form of 3-D cinema involves projecting two images with light waves polarized in perpendicular directions. The polarized lenses act as filters, allowing only one image to reach each eye.

2.6 Optical instruments

Microscopes and telescopes exploit the wave properties of light to create magnified and intensified images.

Optical instruments manipulate light in order to produce a magnified image or to examine a substance more closely. They have enabled landmark discoveries in astronomy, medicine and biology.

- Lenses harness refraction to bend the path of light rays, while carefully shaped mirrors do the same using reflection. Prisms and diffraction gratings use refraction and diffraction, respectively, to disperse light into different colours.

- Telescopes capture light from distant objects. Light rays from stars and planets effectively travel in parallel. By bringing the rays into focus, a telescope creates a virtual image that appears to come from much closer to the observer. By collecting light using a large lens or mirror, a telescope not only magnifies but also creates a much brighter image. Astronomer Galileo Galilei was a telescope pioneer, using an early model to discover the moons orbiting Jupiter.

- In a spectroscope, a telescope is combined with a prism or diffraction grating to produce a spectrum. This can be used to determine the chemical constituents of a substance.

- Microscopes gather diverging light rays from small, nearby objects and use one or more lenses to create a greatly magnified image. In the 17th century, the microscope enabled scientists to see cells and microorganisms for the first time, transforming their ideas of how diseases were transmitted.

Isaac Newton built the first working reflector telescope in 1668.

Simple telescopes

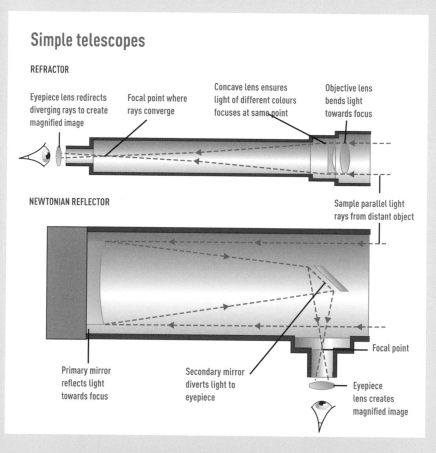

REFRACTOR

Eyepiece lens redirects diverging rays to create magnified image

Focal point where rays converge

Concave lens ensures light of different colours focuses at same point

Objective lens bends light towards focus

NEWTONIAN REFLECTOR

Sample parallel light rays from distant object

Focal point

Primary mirror reflects light towards focus

Secondary mirror diverts light to eyepiece

Eyepiece lens creates magnified image

Telescopes collect light from distant objects and bend it to a focus using either a lens or a mirror. While lens-based telescopes have a simpler design, all large modern astronomical telescopes are reflectors, because a reflecting telescope is much shorter for the same magnification.

2.7 Interferometry

Interference between separated waves is used to calculate otherwise impossible measurements.

Wave interference can be manipulated to provide highly accurate measurements to within 1 nanometre (a billionth of a metre). This is used throughout science and industry and is particularly important for manufacturing electronics.

Interferometers commonly use visible light lasers, although other kinds of electromagnetic waves are also employed. When two waves interact, constructive interference produces bright light, while destructive interference cancels the waves out, leaving darkness. For any situation in between, a pattern of light and dark bands is seen (see Topic 2.4).

In optical interferometry, a single beam of coherent light, such as a laser beam, is split in two by passing it through a semi-silvered mirror. The two beams of light then take different paths. One of the beams travels to a distant mirror and reflects back again. This acts as the norm. The other hits a second, modified, surface and returns, where it recombines with the other beam.

If there is a difference in the distances travelled by each of the two beams, the telltale light and dark bands are produced. Since the wavelength of light is known, it is possible to use this pattern to calculate distance.

Interferometers are often used in industry to scan smooth surfaces to reveal the tiniest bumps.

Astronomers use interferometry to combine signals from different telescopes and mimic the behaviour of much larger instruments.

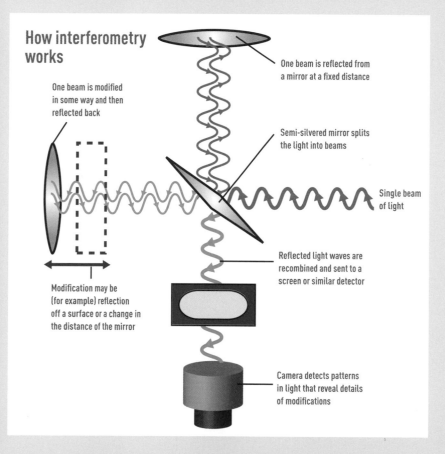

How interferometry works

One beam is modified in some way and then reflected back

One beam is reflected from a mirror at a fixed distance

Semi-silvered mirror splits the light into beams

Single beam of light

Reflected light waves are recombined and sent to a screen or similar detector

Modification may be (for example) reflection off a surface or a change in the distance of the mirror

Camera detects patterns in light that reveal details of modifications

This diagram shows the various processes involved when using a Michelson interferometer. The instrument has a variety of uses, which include very precise distance measurement, the testing of optical components and tuning to observe only very specific wavelengths of light.

2.8 Sound waves and acoustics

Sound waves travel through the air as vibrations, so how do we then make sense of them?

Sound waves are longitudinal. Their waves move in the direction of travel. Sound enables human speech and music, while animals use frequencies we can't hear.

As a sound wave travels through any medium, it makes the particles vibrate back and forth around their natural position. In the air, this creates regions of high and low pressure, which in turn vibrate our eardrums, so allowing us to hear sound.

A human ear can pick up variations amounting to just one-billionth of the ambient atmospheric pressure, at frequencies ranging from 20 to 20,000 hertz. Our brains interpret these different frequencies as pitches.

Variations in sound pressure are so great that a **logarithmic scale** is used. Measured in decibels (dB), an increase of 6 dB signifies a doubling of pressure. The sound pressure level of rustling leaves is 20 dB while the deafening noise of a nearby jet plane is 130 dB.

Whales make use of both ultrasound and infrasound – the first for echolocation, the second for communication over vast distances.

Waves beyond the range of our hearing are referred to as either infrasonic or ultrasonic. **Infrasound** is lower than 20 hertz and is produced by earthquakes and also by animals such as elephants, which use it for communication. **Ultrasound** is a high-pitch sound above 20,000 hertz: it's used in medical scans of unborn babies and also by dolphins and bats for echolocation.

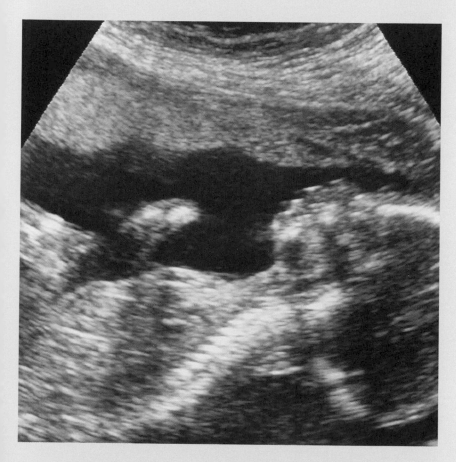

Medical ultrasound imaging applies low-energy, high-frequency waves to body tissues using a device called a sonograph. An image is reconstructed based on the sound-reflecting properties of internal organs and, in the case above, unborn babies.

2.9 The Doppler effect

When the source of a sound wave is moving, its wavelength appears to be stretched or compressed.

Discovered by Christian Doppler in 1842, the **Doppler effect** is a shift in the frequency of waves caused by motion. Not only is it responsible for familiar effects on Earth, but it provides the evidence that our universe is expanding.

You hear the Doppler effect every time an emergency vehicle passes you on a road, its siren blaring. As the siren moves towards you, the wave fronts reach you more quickly than they would if the vehicle was static. The frequency is higher and therefore you hear a higher-pitched sound. Once in retreat, waves arrive more slowly, causing a fall in frequency and pitch, and a longer wavelength.

These shifts affect waves of all kinds. A Doppler radar, for example, bounces microwaves off a moving object and analyses the change in frequency. The police use this kind of radar to determine the speed of cars and can apprehend drivers breaking the limit.

Most galaxies are moving away from the Earth, so the Doppler shift causes them to appear redder than they are.

Sources of light moving towards Earth show a **blueshift** to shorter wavelengths, while those moving away are **redshifted** to longer wavelengths. By measuring the redshifts of galaxies, astronomers came up with an intriguing relationship: the further away galaxies are from Earth, the faster they tend to be moving away from us. This shows that everything in the universe was closer together in the past, and is evidence that it started from a single point in time, known as the big bang.

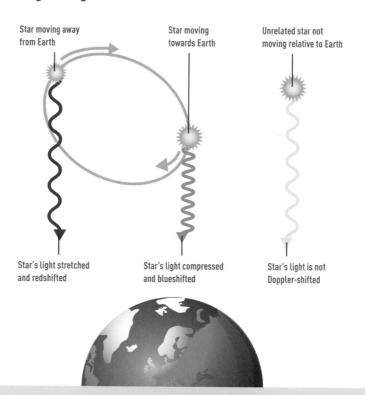

Shifting starlight

Star moving away from Earth

Star moving towards Earth

Unrelated star not moving relative to Earth

Star's light stretched and redshifted

Star's light compressed and blueshifted

Star's light is not Doppler-shifted

Watching for Doppler shifts in starlight can reveal double star systems too closely bound to separate in even the most powerful telescope. The amount of Doppler shift can even reveal the relative masses of the stars involved.

2.10 Modulation

Imposing data signals onto carrier waves allows the transmission of information over great distances.

Soon after the discovery of radio waves in the 19th century (see Topic 4.4), engineers and scientists realized that such waves could cross long distances. This became extremely useful when **modulation** enabled them to carry information. This involves overlaying a continuously varying information signal onto a uniform carrier wave in one of two ways.

- In amplitude modulation (the AM signal on a radio), the strength of the information signal is used to vary the amplitude of a carrier wave with otherwise uniform amplitude.

- In frequency modulation (used in FM radio), the signal strength controls a circuit that varies the frequency of the carrier wave. A radio receiver has a demodulator to extract the information.

The frequency of a wave is far less susceptible to disruption than its amplitude – for example, by atmospheric interference. This is why FM radio broadcasts often sound better than those from AM stations. However, FM radio signals cannot travel as far, and are limited to around 80 km (50 miles).

Amplitude modulation was the earliest method used to transmit the human voice over radio.

In modern times, the move to digital radio enables more information to be carried over longer distances. Digital data consists of a series of zeros and ones (bits). Representing digital data using an analogue wave requires both amplitude and frequency modulation at the same time. Digital modulation is also used in Wi-Fi routers.

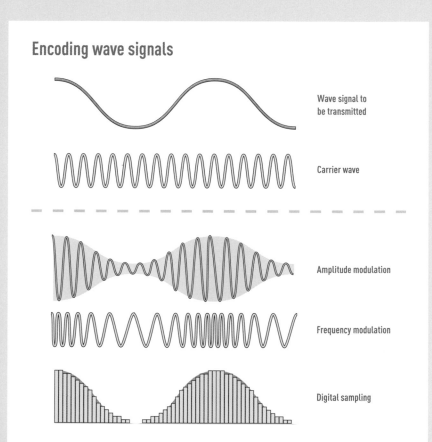

Encoding wave signals

Wave signal to be transmitted

Carrier wave

Amplitude modulation

Frequency modulation

Digital sampling

Three different methods of encoding the same wave-like signal. Amplitude and frequency modulation make use of a carrier wave, while digital encoding measures the wave's value at regular intervals and converts this information into a stream of numbers.

THERMODYNAMICS

Thermodynamics is the study of heat in motion. In the simplest terms, boiling a pan of water by transferring heat from one place to another is a demonstration of thermodynamics. But this field is much broader than that. In some ways this is the most important branch of physics, because it defines the rules of the game for all the others.

The field of thermodynamics developed from studies of steam engines and related devices amid the Industrial Revolution of the 18th and 19th centuries. As such, it proved to be an unusual case of a major technological advance that was put to good use and substantially refined long before its underpinning physical principles were fully identified and understood.

Yet, as scientists like Sadi Carnot in France and Rudolf Clausius in Germany began to formulate the laws

Continues overleaf

of thermodynamics, it became clear that their ideas had implications far beyond the efficiency of steam engines. Heat, after all, is just another form of energy, and the transfer of energy within different systems is the essential prerequisite for every physical process. Therefore the laws of thermodynamics tell us fundamental facts about the behaviour of energy in our universe – not only what we can expect out of apparently unrelated physical systems, but even out of the cosmos when considered as a whole.

The concepts behind thermodynamics can often be confusing, so this chapter starts by looking at the most familiar forms of heat and their sometimes surprising behaviour. We then move on to look at the meanings of heat and temperature, before getting to the heart of the matter – the laws of thermodynamics and their implications for physics as a whole.

Contents

20

30

40

50

3.1 Heat transfer

Heat can be transferred from particle to particle, transported in matter or radiated across space.

Heat is the transfer of energy from a hot body to a colder one. This kinetic energy can be transferred in three ways: conduction, convection and radiation. Thanks to the work of physicists over the last 150 years, the understanding of heat transfer helps us design energy-efficient houses, computers and other appliances.

■ **Conduction** involves energy moving from hot parts of an object to cold ones through the vibration of atoms and molecules. In solids, the molecules are closely packed, so it's relatively easy for them to bump into their neighbours and transfer energy. Some solids are better thermal conductors than others. A saucepan is made of metal because it will warm up quickly.

■ In liquids and gases, atoms and molecules are not so tied down and so have more kinetic energy. The movement of high-energy molecules increases the volume in one part of a liquid, making it less dense. This rises above denser parts, replacing colder liquid and setting up a **convection** current. If you warm soup on a hob, the soup at the bottom of the saucepan warms up first, then rises to the surface.

All the heat reaching Earth from the Sun is transferred by means of radiation.

■ Every object emits **thermal radiation**, which is just light. Radiation can travel through the vacuum of space, which is how radiation from the Sun keeps us warm.

The Sun demonstrates all three methods of heat transfer – energy escapes from the core through conduction and radiation, before being absorbed to heat up gas in the outer, convective layers. Finally, radiation escapes into space at the Sun's surface.

3.2 Heat capacity

The amount of energy required to raise the temperature of something depends on the substance from which it is made.

In order to heat something up, it's useful to know how much energy you're going to need. Fortunately, there is a method for comparing how easily you can raise the temperature of different materials, and this is known as heat capacity. It also tells us why water is so good for cooling things down.

Scientists talk about **specific heat**, which is the amount of energy required to raise the temperature of a unit mass of, say, 1 kg (2.2 lb) of a material by 1 degree Celsius (equivalent to a change of 1.8 °F). Heat capacity is given as joules per kilogram per kelvin (K), expressed as J/kg/K. This is because a rise of 1K is equivalent to a rise of 1 °C and kelvin is one of the standard (SI) units of measurement used in science (see Topic 3.5).

Most common substances have a heat capacity in the range of 100 to about 1,000 J/kg/K, but water is much higher at 4,200 J/kg/K. Water is also a good thermal conductor (see Topic 3.1), and this combination makes it invaluable as a coolant. When poured over a hot baking tray, for example, cold water takes heat away from the hot tray. However, the high heat capacity of water means it barely changes temperature itself, which is why it is commonly used as a coolant for braking systems.

Among the elements, hydrogen has the highest heat capacity of all, while the heavy gas radon has the lowest.

The heat capacity of different materials

Wood has a high heat capacity and increases temperature slowly

Metals have a low heat capacity and increase temperature rapidly

As the water in the pan begins to boil, the heat circulates by convection, warming the two spoons

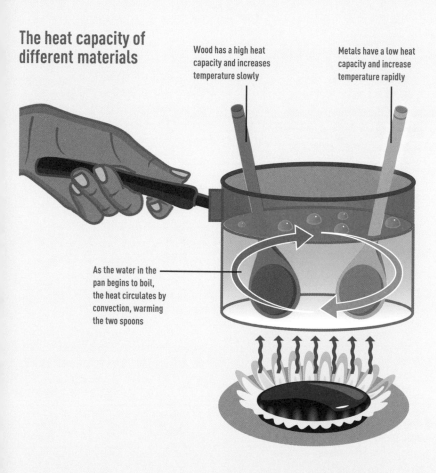

The difference in behaviour between a wooden and a metal spoon when placed in boiling water demonstrates their different heat capacities.

3.3 Latent heat

Breaking the bonds in a substance takes energy, while forming new bonds releases energy.

Latent heat is the energy associated with a phase change – for example, a solid changing into a liquid. It involves a material absorbing energy without a rise in temperature.

Think of ice melting to water. Water drips from the melting ice while maintaining its temperature of 0 °C (32 °F). Energy is being absorbed, but the ice isn't getting any hotter. Instead of giving the atoms or molecules kinetic energy – that is, making them vibrate more vigorously – the energy is absorbed by the bonds that tie the molecules together. Beyond a certain threshold, these bonds break, and so the ice melts. A similar process occurs when water evaporates. Boiling water stays at the same temperature – 100 °C (212 °F) – while energy is absorbed as part of the vaporization process.

Latent heat associated with melting is called the latent heat of fusion. When associated with evaporation it's called the latent heat of vaporization. Both are measured in joules per unit of mass. The latent heat of fusion for water at 0 °C (32 °F) is 334 joules per gram, while its latent heat of vaporization at 100 °C (212 °F) is 2,260 joules per gram.

Just as energy is absorbed during the melting and evaporation processes, it is released when water condenses and when it freezes (see opposite).

Large quantities of latent heat are absorbed as water evaporates off Earth's oceans. Heat is released again, when the water condenses in the upper atmosphere.

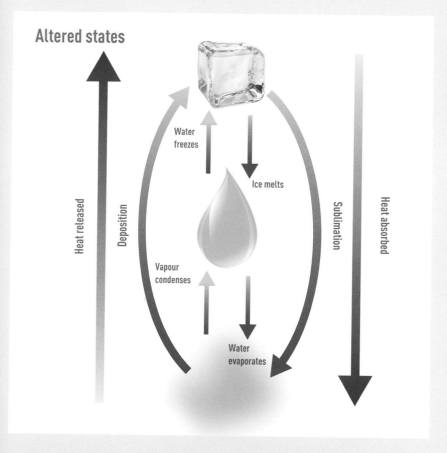

In the right conditions some substances can change state directly between a solid and a gas (sublimation) or a liquid and a solid (deposition), and vice versa. They absorb or release large amounts of heat as they do so.

3.4 Measuring temperature

Measuring temperature comes down to the average kinetic energy of a material's atoms or molecules.

When two substances come into contact with one another, how can you tell which is the hottest and, therefore, which way the energy will flow? This is where the concept of temperature comes in, because temperature measures the average kinetic energy of the atoms or molecules that make up a substance.

For temperature to be useful, it needs to be compared to something else – a scale of some kind that is defined by fixed points. Typically, these fixed points are the boiling and freezing points of water. The fixed points are given particular values and the scale is split into divisions called degrees. Familiar temperature scales include Fahrenheit, in which the freezing point of water is 32 °F and the boiling point of water is 212 °F. The difference between the two, 180 degrees, was deemed an awkward number and so the Celsius scale was introduced with 100 degrees between the fixed points.

Mercury expands quite slowly over common temperature ranges – allowing thermometers to be kept small.

In practice, in order to measure the temperature of something, you need a substance that expands and contracts respectively when heated and cooled. The earliest thermometers used columns of air and even wine, but Gabriel Fahrenheit settled on mercury. More modern electronic thermometers rely on other temperature-sensitive properties, such as conductivity or resistance.

How a thermometer works

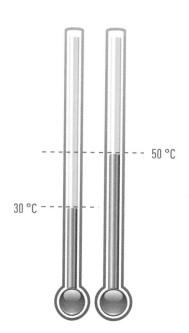

50 °C

30 °C

Mercury expands steadily as
temperature rises, forcing the
liquid column upwards

Comparing temperature scales

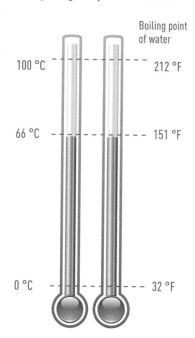

Boiling point
of water

100 °C — — — — — — — 212 °F

66 °C — — — — — — — 151 °F

0 °C — — — — — — — 32 °F

Identical Celsius and Fahrenheit
thermometers will have the same height
of mercury at the same temperature

A liquid thermometer relies on the expansion of mercury along a narrow tube. Graduated
marks along the tube can show any temperature scale, but if the tubes are identical, the
level of mercury remains the same.

3.5 Absolute zero

How cold is the coldest possible temperature? The Kelvin scale reveals all.

When physicists began to develop a greater understanding of the energy of atoms and molecules, they also developed a new concept – absolute zero. In the process, they created a scale that is widely used in scientific measurements.

Fahrenheit and Celsius have one disadvantage. When the German physicist Daniel Fahrenheit devised his scale, 0 °F was the lowest temperature he could create in his lab using a mixture of salt, ice and water. But temperatures can be much colder than this, which introduces negative values. In the Arctic in winter, for example, it gets down to a chilly -40 °C (-40 °F).

Even colder temperatures than this are possible, however, and that is where the new scale comes to the fore. It measures the average kinetic energy of molecules, and the lowest conceivable temperature occurs when molecules stop moving altogether. This happens at –459.67 °F (–273.15 °C), otherwise known as absolute zero.

Half a billionth of a degree above absolute zero is the lowest temperature achieved in a laboratory.

In 1848, British physicist William Thomson (Lord Kelvin) argued that absolute zero could be used as the basis for an absolute temperature scale. Today, this scale is defined in terms of units called kelvins (K), where a change of 1 K is equivalent to a change of 1 °C (1.8 °F). According to this scale, the freezing point of water is 273.15 K.

At low enough temperatures, even the gases of our atmosphere condense
into liquid form. Liquid nitrogen has a boiling point of −195.8 °C (−320 °F/77 K)
and rapidly evaporates back into the air when released from a container.

3.6 Heat engines and pumps

Heat engines transfer energy in order to drive mechanical work. Heat pumps apply mechanical work to extract heat.

Once scientists began to explore the principles of thermodynamics, their newfound knowledge was put to good use designing engines to do a range of tasks.

A heat engine is a system that performs mechanical work via the transfer of energy. A typical heat engine converts heat into mechanical force by changing the physical properties of a working fluid, such as water. It uses principles of mechanics such as the gas laws (see Topic 1.9) to convert the working fluid into useful movement. Some energy is always lost in the process, an effect known as entropy (see Topic 3.7).

A heat engine is often driven by a change in the phase of the working fluid. A steam engine uses the transition of water between liquid and a vapour, for example. Steam occupies more space than liquid water, which raises the internal pressure in a chamber to push on a piston.

An internal combustion engine, in contrast, generates motion without a phase change and uses only the expansion and contraction of a hot vapour.

The rules of heat engines were established more than 100 years after the first steam engines put the principles to work.

A heat pump works in reverse, harnessing mechanical forces to change the phase of a working fluid as it moves around a system. The heat pump in a refrigerator evaporates a fluid to absorb heat from the interior. The gas is then compressed back to a fluid to release heat to the outside world.

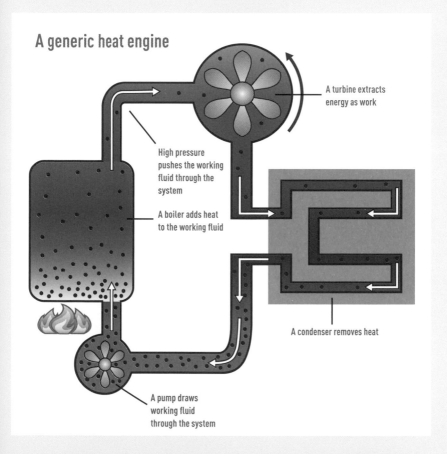

A generic heat engine

A turbine extracts energy as work

High pressure pushes the working fluid through the system

A boiler adds heat to the working fluid

A condenser removes heat

A pump draws working fluid through the system

In this example, a heat source raises the temperature of a working fluid, and a condenser lowers it again. The difference between the two temperatures sets the engine's theoretical efficiency. Thanks to entropy, however, a heat engine can never be 100 per cent efficient (see Topic 3.7).

3.7 Enthalpy and entropy

Just how much of the total energy in a system is put to good work?

The distribution of energy through any thermodynamic system can be described in terms of two related properties – enthalpy and entropy.

A system's **enthalpy** (usually indicated by an H) is the total energy the system contains, including the energy supplied in order to make it work. It is not possible to measure enthalpy directly, but knowing the change in enthalpy is extremely useful. Under conditions of constant pressure, for example, the change in enthalpy is the heat added to the system.

Entropy (S) is the amount of thermal energy in the system that can't be utilized for doing work. Heat engines rely on exploiting uneven distributions of thermal energy between different parts of a system to do work (see page 73). They increase the entropy of a working fluid in some places and decrease it in others, converting it from a disordered hot gas to a more orderly cooler liquid, for example. But the cycle of a heat engine won't repeat endlessly on its own, because some energy is always lost in heating up the engine components and the surrounding air.

Entropy itself is often considered an indicator of the disorder in a system – if you spill some milk from a glass, the entropy of the system increases. According to the second law of thermodynamics, it cannot be reversed without putting energy in – in this case sponging up the mess.

The word enthalpy comes from the Greek 'to add heat'.

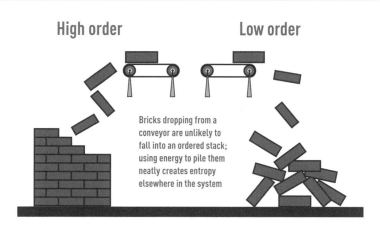

High order   Low order

Bricks dropping from a conveyor are unlikely to fall into an ordered stack; using energy to pile them neatly creates entropy elsewhere in the system

Even within the most efficient systems, hot and cold particles tend to mix together and even out over time

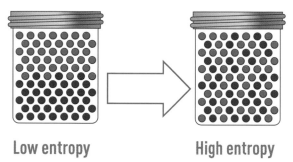

Low entropy   **High entropy**

Entropy's tendency to increase means that neatly ordered systems do not tend to occur of their own accord or persist in isolation. In general, they can only be created by using energy from the surrounding environment, thereby increasing its entropy.

3.8 Laws of thermodynamics

Four laws describe the behaviour of energy in closed systems and across the universe as a whole.

Since the 19th century, scientists have developed four laws of thermodynamics to describe the properties of energy and its conversion into different forms.

■ **The first law** states that the internal energy of a closed system changes when heat is transferred in or out, either by work being done to the system or the system doing work. This rules out hypothetical perpetual motion machines that perform work without external energy input (see opposite).

■ **The second law**, summarized as entropy increases, states that entropy in an isolated system never decreases, and that a system will evolve towards thermodynamic equilibrium (maximum entropy) without the application of external work.

■ **The third law** builds on the second law, stating that entropy in a system only approaches a constant value when its temperature approaches absolute zero.

■ **The zeroth law**, which was formalized after the others, says that if two objects are in thermal equilibrium with a third so that no heat flows between them, then they must also be in equilibrium with each other.

These laws apply not just to steam engines, but to almost every device, including modern-day computers.

Ginsberg's theorem restates the first three laws of thermodynamics: you can't win; you can't break even; and you can't even get out of the game.

Perpetual motion

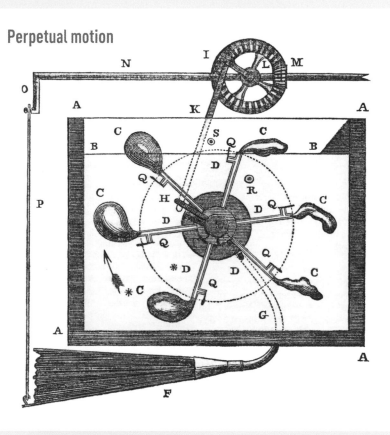

Before the laws of thermodynamics were discovered, many inventors pursued dreams of perpetual motion. They imagined devices like the one above, with the aim of doing mechanical work without using up an initial input of energy.

3.9 Maxwell's demon

An intriguing thought experiment attempts to find a way around the inevitable loss of useful energy to entropy.

When physicists devise laws, they test them by performing various experiments. Sometimes scientists use thought experiments (experiments carried out in thought alone) to test an idea to the limit.

The 19th-century physicist James Clerk Maxwell came up with a thought experiment testing the second law of thermodynamics (see Topic 3.8). He wondered if it was possible to lower entropy without doing any work. He imagined a chamber filled with a gas at a uniform temperature, divided by a barrier containing a door that could be opened and closed.

Although the gas is at a uniform temperature, we know that temperature only ever refers to the average energy of the molecules (see Topic 3.4). Some are faster moving with more energy; others have less. What if a demon guarded the door? He could let high-speed molecules through from left to right but block their return. Conversely, he could let only slow-moving molecules pass from right to left. The result is that the system gradually becomes unbalanced, with one hot side and one cold side.

Maxwell described his door-opening agent as a finite being – it was actually Lord Kelvin who introduced the idea of the demon.

But has entropy really been lowered? Actually, you can argue that work has indeed been done – by the demon who is operating the door. The demon is generating entropy, which more than matches the apparent decrease in entropy caused by the gas becoming more ordered.

How the experiment works

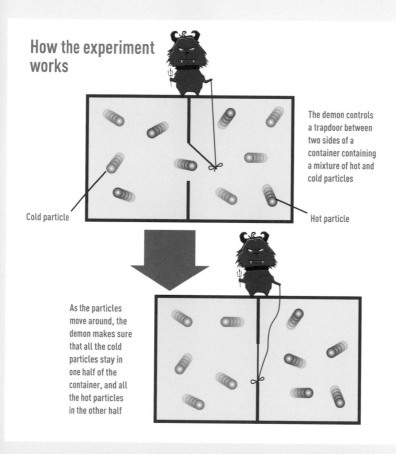

The demon controls a trapdoor between two sides of a container containing a mixture of hot and cold particles

Cold particle

Hot particle

As the particles move around, the demon makes sure that all the cold particles stay in one half of the container, and all the hot particles in the other half

Seen from one perspective, Maxwell's demon is decreasing the box's entropy. In reality, he is part of the system with the box, and the work he does inevitably increases overall entropy.

3.10 Heat death and the arrow of time

Thermodynamics is the only area of physics that defines a direction for time in the universe.

The second law of thermodynamics (see Topic 3.8) gives us the direction in which time is flowing. In doing so, it has fatal consequences for the future of our universe.

Most laws of physics are time invariant. This means they work even if the flow of time is reversed. Yet the second law of thermodynamics says that the entropy of a system – that is, the amount of disorder in its molecules and the amount of thermal energy that cannot be harnessed for work – increases over time unless more energy is put in. This gives laws of physics a so-called **arrow of time**. The fragments of an eggshell cannot reassemble themselves once broken, just as the mixed hot and cold air molecules of Maxwell's demon (see Topic 3.9) are unlikely to separate on their own.

Understanding the forward direction of time is vital to understanding the evolution of the universe itself, from the big bang to the present day. But the second law also has a sting in the tale for future inhabitants of the cosmos. The reason for this is that our own universe is a closed system and is subject to a gradual increase in entropy, although the influence of gravity can help to hold it at bay. As successive generations of stars live and die, the universe will gradually get very slightly warmer. As thermal equilibrium is approached, new stars will cease to form. Ultimately, even matter itself may begin to disintegrate in a phenomenon known as the heat death of the universe.

The arrow of time concept was coined by British astronomer Arthur Eddington in 1928.

The arrow of time explains why we can immediately interpret the events
in this picture, and do not mistakenly imagine a situation in which the
fragments of coffee cup would leap upwards and reassemble themselves.

ELECTROMAGNETIC RADIATION

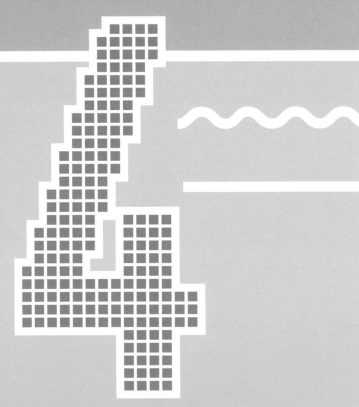

Electromagnetic radiation is a wave like no other – a form of energy that can travel through space without the need for a supporting medium. It can exhibit the properties of both waves and particles and, amazingly, it sets the ultimate speed limit of the universe.

Depending on wavelength and frequency, electromagnetic radiation can exhibit a huge range of properties and be put to work in a wide variety of applications. Visible light is its most familiar form, but there are many others.

At one end of the spectrum are long-wavelength radio waves. These pass unaffected through most obstacles, which makes them ideal for carrying broadcast signals. At the opposite end of the spectrum sit high-frequency gamma rays. These combine their penetrating power with potentially dangerous amounts of energy.

Continues overleaf

In between lie useful microwaves, ubiquitous infrared (produced by almost every object in the universe), energetic ultraviolet and the X-ray, with its diverse range of both medical and scientific applications.

The first half of this chapter looks at the nature of light itself. The various topics consider the characteristics, sources and applications of the different types of electromagnetic radiation one by one. The second half of the chapter explores some further aspects of light and other radiations, including the ways in which light is generated. These range from incandescent surfaces to faster-than-light particles and devices such as lasers.

Contents

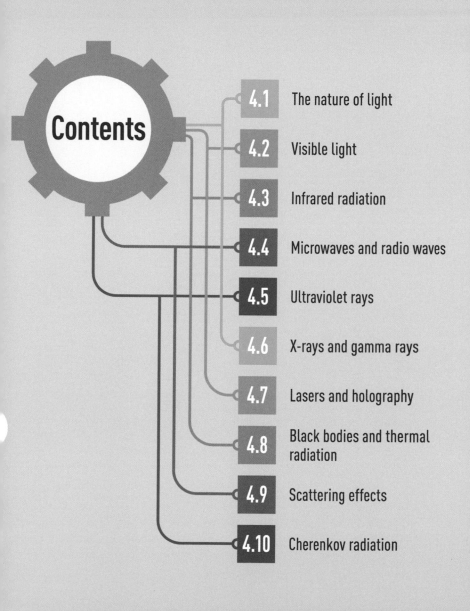

4.1 The nature of light

It took more than 300 years to discover the true nature of light and how it travels through a vacuum.

The exact nature of light was a puzzle for many centuries. Initially, scientists thought it was made of particles. Later, they thought it behaved more like waves. It wasn't until the 20th century that light was shown to have properties of both.

Robert Hooke and Christiaan Huygens advanced arguments for light being a wave in the late 1600s, while it was Isaac Newton who later proposed the idea of tiny particles. In 1704, he summarized the evidence in his work *Opticks*. He pointed out that light always travels in a straight line and suggested that reflection is explained by particles bouncing off a surface.

Newton's view held until around 1800, when Thomas Young pushed the argument back in favour of wave theory, with his double-slit light experiment (see Topic 2.4). The pattern of light and dark bands produced on a screen could be explained by the constructive and destructive interference of waves.

Then, in 1865, James Clerk Maxwell calculated that visible light was a form of electromagnetic wave (see opposite). Such waves could have a huge range of wavelengths but always travelled at the same speed in a vacuum – 300,000 km/s (186,000 miles per second).

The vast majority of bodies are visible only in reflected light, but almost everything emits some form of lower-energy radiation.

Finally, in 1905, Albert Einstein's work on the photoelectric effect introduced the idea of **wave-particle duality** (see Topic 8.1). Light is now often described as packets that display wave-like behaviour.

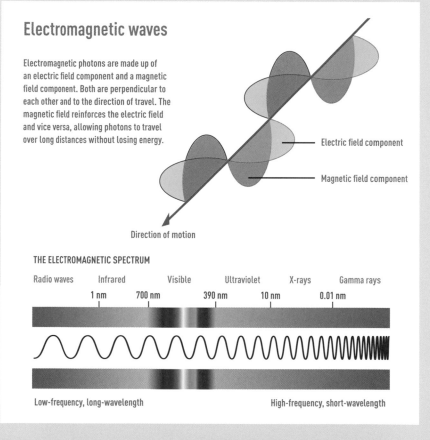

Electromagnetic waves

Electromagnetic photons are made up of an electric field component and a magnetic field component. Both are perpendicular to each other and to the direction of travel. The magnetic field reinforces the electric field and vice versa, allowing photons to travel over long distances without losing energy.

Electric field component

Magnetic field component

Direction of motion

THE ELECTROMAGNETIC SPECTRUM

Radio waves	Infrared	Visible	Ultraviolet	X-rays	Gamma rays
	1 nm 700 nm	390 nm	10 nm	0.01 nm	

Low-frequency, long-wavelength

High-frequency, short-wavelength

Visible light represents just a small range of wavelengths within a much larger spectrum that is measured in nanometres (nm; billionths of a metre). We consider visible light special because our eyes have evolved to detect and interpret the world this way.

4.2 Visible light

Our eyes detect radiation in just one tiny part of the electromagnetic spectrum, processing different wavelengths as different colours.

The region of the electromagnetic spectrum that we know as visible light encompasses the wavelengths we can see with the naked eye. Different wavelengths correspond to different colours.

A narrow window in Earth's atmosphere allows radiation with a specific range of wavelengths to pass through from space unhindered. These wavelengths are between 390 and 700 nanometres (nm; billionths of a metre, see page 87). The shortest of these rays are coloured violet and the longest are red. Human eyes have evolved to be sensitive to these wavelengths.

The major source of visible light is the Sun. When its light reflects off an object, it makes the object visible to us. Objects reflect wavelengths differently – a lemon generally reflects yellow wavelengths, for example. This property gives all objects their distinctive colours.

Until the time of Newton, many people believed that contamination in glass or water produced the rainbow colours.

The Sun emits a broad range, or **continuum**, of wavelengths, as do other bodies depending on their surface temperature (see Topic 4.8). As perceived by the human eye and brain, these wavelengths blend together as white. In 1666, Isaac Newton used a prism to split pure sunlight into its constituent colours (see opposite) – an experiment that also explained the meteorological phenomenon of rainbows.

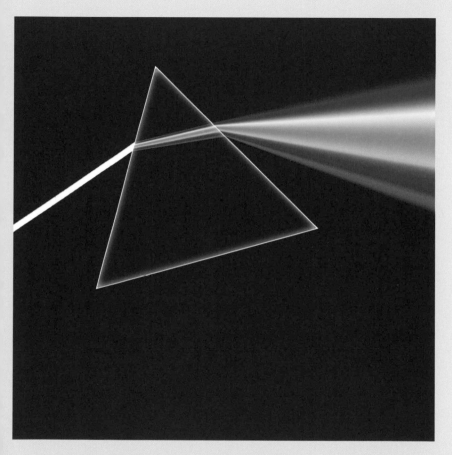

A prism splits light into different colours because light waves diffract by different amounts
depending on colour and wavelength (see Topic 2.3). Newton's famous experiment involved
not just splitting a beam of light apart, but also using refraction to reunite its colours into white.

4.3 Infrared radiation

Rays with longer wavelengths than visible light are felt as invisible heat radiation.

Isaac Newton's use of a prism to split sunlight into its constituent colours (see Topic 4.2) marked the start of many investigations into the nature of light. One experiment led to the discovery of infrared radiation.

Astronomer William Herschel wanted to know whether the colours of the spectrum had different temperatures. In 1800, he placed thermometers on the colours produced by a prism. He noticed that the temperatures rose from the blue end of the spectrum to the red end. A thermometer placed beyond the red end, where there was no visible light, recorded an even higher temperature. Herschel had discovered infrared.

Infrared has wavelengths between 750 nm (see page 87) and 1 mm. A substantial number of shorter-wavelength, near-infrared rays from the Sun make it through Earth's atmosphere, where their energy is felt as heat.

Every object in the universe emits infrared radiation – how much depends on its temperature, a relationship that is described by the black body curve shown on page 101.

Shorter-wavelength, near-infrared rays do not transfer much heat, making them ideal for use in devices such as television remote controls.

Infrared has a variety of applications. Infrared telescopes study galaxies too faint to see using normal telescopes; satellites in orbit use infrared to monitor the health of crops; and rescue teams employ thermal-imaging cameras to detect the heat given off by survivors buried by an earthquake.

An infrared camera can be used to monitor Earth from space – in this satellite image of Italy's Mount Etna volcano, recently erupted lava still has a strong infrared signature (coloured red), while surrounding snowfields appear blue and forests green.

4.4 Microwaves and radio waves

The longest-wavelength electromagnetic waves are ideally suited to transmitting information and even power.

Beyond the infrared region on the electromagnetic spectrum, radiation has even longer wavelengths. These microwaves and radio waves are vital for communications.

The presence of these longer waves was predicted by James Clerk Maxwell, as a direct consequence of his new wave theory (see Topic 4.1). In 1887, such waves were generated in a laboratory by Heinrich Hertz. He showed that an oscillating current in one antenna could create an oscillating current in a second antennae across the room. Initially called Hertzian waves, we know them today as radio waves. In radio broadcasting, the most efficient antennae usually take the form of a radio mast – usually a thin dipole conductor consisting of two horizontal rods (see opposite).

Microwaves have wavelengths between 1 mm and 1 m (3 ft). Contrary to the suggestion made by their name, this isn't very short; 'micro' simply means the waves are small compared to radio waves. They are easier to focus into beams than radio waves, which makes them suitable for directional communications, such as TV and data beamed to Earth from satellites. They can also pass through food and transfer energy to molecules of water – properties exploited by the microwave oven. Generating microwaves requires specialized equipment in which electrons oscillate under the influence of an electromagnetic field.

The afterglow of the big bang at the start of the universe is detected as microwaves.

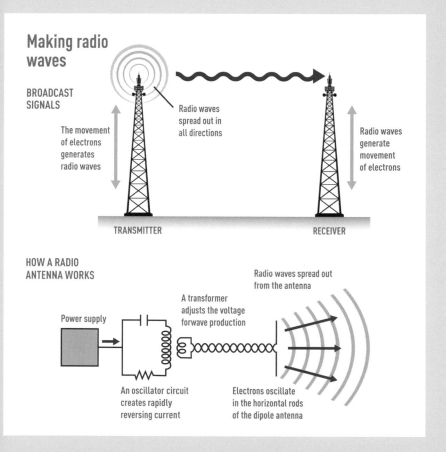

Making radio waves

BROADCAST
SIGNALS

The movement
of electrons
generates
radio waves

Radio waves
spread out in
all directions

Radio waves
generate
movement
of electrons

TRANSMITTER

RECEIVER

HOW A RADIO
ANTENNA WORKS

Radio waves spread out
from the antenna

A transformer
adjusts the voltage
forwave production

Power supply

An oscillator circuit
creates rapidly
reversing current

Electrons oscillate
in the horizontal rods
of the dipole antenna

Radio waves are generated by electrically charged particles, called
electrons, moving back and forth in a conducting wire (see Topic
6.1), called an antenna.

4.5 Ultraviolet rays

Radiation with shorter wavelengths than visible light has many uses. It is also potentially dangerous to living cells.

Infrared, microwaves and radio waves all have longer wavelengths than visible light. Radiation with wavelengths just shorter than visible light is called ultraviolet.

Ultraviolet ranges in wavelength from 400 nm down to 10 nm. It was discovered in 1801 by German physicist Johann Wilhelm Ritter, who had heard about Herschel's discovery of infrared (see Topic 4.3) the year before. In an experiment using silver chloride salt, which turns black when exposed to light, Ritter found that the reaction continued to happen beyond the violet end of the spectrum. He'd found ultraviolet, known initially as chemical rays.

The Sun does not produce as much ultraviolet as visible light, but ultraviolet carries more energy. When it strikes human skin it can damage DNA in the cells, which can lead to cancer-causing mutations. Ozone, an unusual form of oxygen molecule high in Earth's atmosphere, blocks almost all the highest-energy ultraviolet radiation (UV-C), but lets through a little UV-B and most UV-A wavelengths.

Ultraviolet light is relatively easy to produce artificially and has many applications. It can be used to kill harmful bacteria in hospitals, to trigger chemical reactions in manufacturing and to cause some substances to fluoresce – useful for detecting forged bank notes or for marking property with a signature that is invisible to the naked eye.

Some flowers have ultraviolet markings that help guide bees to their nectar.

Ultraviolet protection

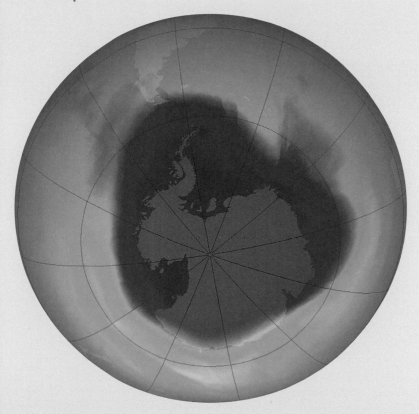

A satellite map shows (in purple) the hole in Earth's ozone layer above the Antarctic at its greatest extent, in 2008. This layer, 20 to 30 km (12 to 19 miles) above Earth's surface, is rich in ozone molecules. They interact with, and absorb, ultraviolet light, reducing the amount that reaches Earth.

4.6 X-rays and gamma rays

High-frequency, electromagnetic waves can pass through many substances unaltered. This makes them difficult to detect.

The forms of electromagnetic radiation with the highest energy and shortest wavelengths are X-rays and gamma rays. Both have important uses in medicine, thanks to their ability to penetrate the human body.

X-rays have wavelengths from 10 nm down to 0.01 nm and were discovered by Wilhelm Röntgen in 1895. He found that an electrical discharge tube, a device similar to a cathode-ray tube, caused a nearby object to glow, even though the tube was completely covered. He dubbed this invisible, unknown form of energy 'X-rays'.

Scientists soon discovered that his rays pass easily through human tissue but not bone. This made them ideal for medical imaging, using X-ray photographs to reveal broken bones and tooth decay. It was many more years before scientists understood the dangerous ability of X-rays to trigger cell mutations. Today, exposure is limited wherever possible.

From the 1920s to the 1950s, X-ray machines were widely used in shoe stores, supposedly to ensure the best fit.

Higher-frequency gamma rays are created all of the time on Earth – in small doses via the process of radioactive decay (see Topic 7.1). Gamma rays pack an even greater punch than X-rays. It takes a 1.8-m (6-ft) thickness of concrete to create an effective shield. If gamma rays pass through the human body, they damage tissue and DNA. This same property means they can be used to kill cancer cells and shrink tumours.

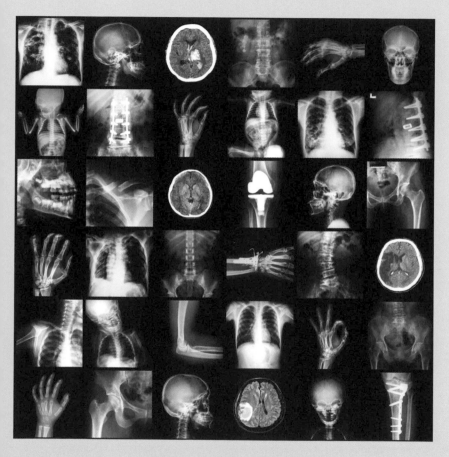

X-rays today are used largely for medical and scientific research
purposes. The potential harm from repeated exposure became clear early
on, but this did little to prevent their overuse in the early 20th century.

4.7 Lasers and holography

Consisting of light waves that keep in step with one another, a laser is an intense light source and a powerful tool.

The first laser was built as recently as 1960, inspired by physics Einstein had developed more than 40 years earlier. Today, lasers have a wide variety of uses, from playing DVDs through barcode scanning to tattoo removal and eye surgery.

The word 'laser' is an acronym for light amplification by stimulated emission of radiation. At its heart is a suitable **lasing medium** – it can be a solid, a liquid or a gas. When the medium's atoms are stimulated, they emit photons of light that cause other atoms to emit photons as well, so creating a cascade. The photons all have the same frequency (colour) and travel in the same direction. In early lasers, this cascade effect is started by an intense burst of light from a flash tube, and enhanced by mirrors at each end of a laser-making device, which reflect the photons. One mirror is partially silvered, allowing a narrow beam to escape. The emitted photons are **coherent** – they move in step – and the beam remains tight over a long distance. This means that it can be exploited for everything from precision cutting to targeting systems for weapons.

The strength of early lasers was measured in Gillettes, based on the number of razor blades the beam could burn through.

Holography exploits the interference and diffraction of laser beams. Three-dimensional **holograms** are created when a photographic plate records the light field produced when a laser interacts with a solid object. Holography can also be used to store huge amounts of computer data, in what may become an everyday technology in the future.

How a laser works

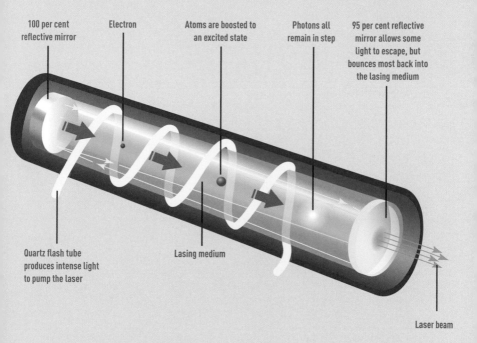

100 per cent reflective mirror

Electron

Atoms are boosted to an excited state

Photons all remain in step

95 per cent reflective mirror allows some light to escape, but bounces most back into the lasing medium

Quartz flash tube produces intense light to pump the laser

Lasing medium

Laser beam

Inside a laser, collisions between photons and excited atoms cause the atoms to lose energy and release light, rather than this happening naturally. The result is an intense beam of identical photons bouncing back and forth in the lasing medium, triggering more and more emission.

4.8 Black bodies and thermal radiation

It is possible to assess the surface temperature of a distant star, simply by the colour it emits.

Any object with a finite temperature emits electromagnetic radiation all of the time. The wavelengths given off, and the intensity of the radiation emitted, depend on that object's surface temperature.

If you watch a metal bar heated in a furnace, you will see it emit different wavelengths over time. It first glows red then orange, moving through yellow to white, and finally blue. Even before you see the metal glow visibly, it will be emitting infrared that you will feel if you put your hand close to it. If heated even more, emissions shift beyond the blue end of the spectrum into the high-energy, short-wavelength ultraviolet range.

During the mid-19th century, physicists developed a means of predicting the spread of radiation from a hot object by devising a **black body**. This was a theoretical, ideal object with a perfect, light-absorbing, non-reflective surface. They used it to demonstrate the relationship between surface temperature and wavelengths emitted (see opposite). While the concept seems abstract, it's a surprisingly accurate model for the behaviour of objects ranging from stars to incandescent lamp filaments. The relationship can be used to work out the surface temperatures of stars from their colours. For example, the star Betelgeuse appears red and has a surface temperature of 3,500 K; Rigel, a star that appears white, has a surface temperature of 11,000 K.

Our Sun has a black body temperature of about 5,430 °C (9,800 °F).

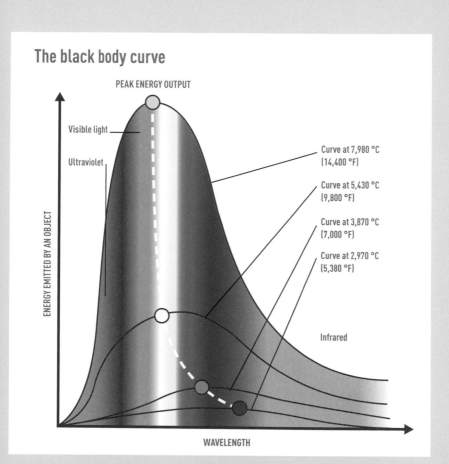

The black body curve

PEAK ENERGY OUTPUT

Visible light

Ultraviolet

ENERGY EMITTED BY AN OBJECT

Curve at 7,980 °C
(14,400 °F)

Curve at 5,430 °C
(9,800 °F)

Curve at 3,870 °C
(7,000 °F)

Curve at 2,970 °C
(5,380 °F)

Infrared

WAVELENGTH

The spread of radiations from a black body follows a distribution defined by the Stefan-Boltzmann law, and is often known as the object's black body curve. This chart shows black body curves corresponding to a variety of temperatures.

4.9 Scattering effects

Light waves interact with particles in Earth's atmosphere to produce blue skies, rainbows and more.

When objects are scattered, like pool balls in a collision, they move off in different directions. Light behaves like a particle so it, too, can scatter, giving rise to a range of natural phenomena from blue skies to rainbows. Scattering can be either **elastic** or **inelastic**, depending on whether individual particles of light – photons – retain or lose energy after an interaction.

- Photons retain their energy in collisions with particles that are electrically neutral and small compared to the photon's own wavelength, such as the atoms and molecules in our atmosphere. Such elastic collisions result in **Rayleigh scattering**. The original photon affects the particle's internal electric field, causing it to oscillate and emit light of its own in all directions. The scattering effect is strongest for shorter, bluer wavelengths. As a result the scattering of sunlight, for example, makes the sky blue and direct light from the Sun more yellow than it would otherwise be (thanks to the removal of its blue component).

- **Compton scattering** is an inelastic effect that occurs when light interacts with charged particles, such as electrons. The particle gains energy, while the photon loses it and consequently emerges with a lower frequency and longer wavelength.

Some 25 per cent of incoming sunlight is scattered by Earth's atmosphere. Two-thirds of this ends up as diffuse sky glow.

Scattered light

RAYLEIGH SCATTERING

Rays from the Sun

Atmosphere

Interaction with small uncharged particles scatters short wavelengths in all directions, turning the sky blue

The original light beam emerges with less energy and redder in colour

COMPTON SCATTERING

A charged electron steals energy from the photon and is deflected in the process

An incoming short-wavelength photon

The photon is deflected and has a longer wavelength due to loss of energy

Rayleigh and Compton scattering are the main elastic and inelastic scattering processes. Compton scattering is particularly important in medicine since it is used to predict the effects of gamma rays on patients during radiotherapy.

4.10 Cherenkov radiation

When particles in a medium move faster than the speed of light, they produce the optical equivalent of a sonic boom.

Light travels through space at 300,000 km/s (186,000 miles per second), but is slowed down in water, ice, air and other media. In water, for example, it only reaches around 225,300 km/s (140,000 miles per second). At these slow speeds, electrically charged subatomic particles can sometimes travel faster than light.

The situation for a so-called **superluminal** particle can be compared to that of a supersonic plane. When an aircraft travels faster than sound, it breaks the sound barrier and you hear a sonic boom. The optical equivalent is seen as the blue glow of Cherenkov radiation.

The radiation is released because a particle's passage through the medium rearranges the electromagnetic fields around it, causing molecules to become polarized (see Topic 2.5) before reverting to their original state. Excess energy released as the molecules relax takes the form of visible light emitted along what's known as a **shock front**.

Cherenkov radiation is routinely produced when fast-moving particles from nuclear reactor cores leak out into the coolant. Similarly, high-speed particles known as cosmic rays can generate Cherenkov radiation as they rain down on Earth.

British physicist Oliver Heaviside predicted the existence of Cherenkov radiation 70 years before its discovery.

Cherenkov radiation is best known as the blue haze associated with nuclear reactor cores. Scientists are fast finding applications for it – not only in physics, but in fields such as medical imaging.

STRUCTURE OF MATTER

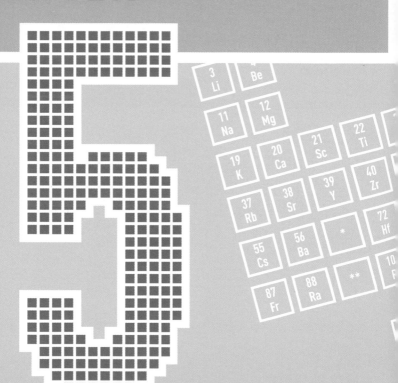

What is matter? This question has puzzled philosophers and scientists since ancient times. It also lies at the heart of modern physics. Several of the later chapters in this book are devoted to addressing the question in detail. They focus on the various strange and counter-intuitive things that seem to happen on the very smallest scale. However, before reading about those, it is important to gain an understanding of the fundamental properties of matter – the physical characteristics and chemical behaviours that define atoms, molecules and elements.

The idea that everything in the universe is made from some ultimately indivisible substance is often attributed to the Greek philosopher Democritus. The concept also enjoyed moments of popularity as a philosophical theme elsewhere in the ancient world.

Continues overleaf

Today's scientific atomic theory originated in the early 19th century with the work of chemist John Dalton. He wondered why his chemical reactions often used up fairly simple, fixed ratios of different substances. His conclusion was that everything must consist of individual atoms of different substances that were grouped together, often in small numbers. This simple idea led to a revolution in chemistry during the next century, and resulted in the discovery of dozens of distinct chemical elements.

A true understanding of the exact nature of the atoms that make up the different elements had to wait until the turn of the 20th century. A series of findings showed that atoms were themselves made of smaller particles – protons, neutrons and electrons. New theories suggested that, ultimately, the behaviour of a particular atom is determined by the number and configuration of these subatomic particles.

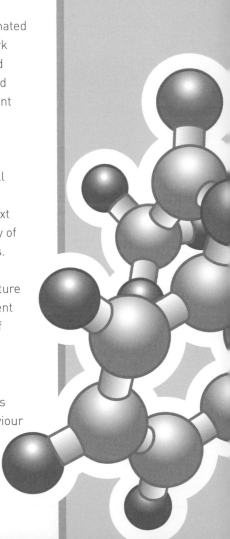

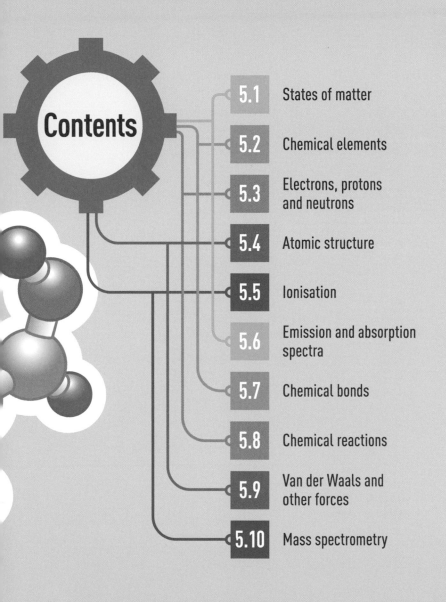

Contents

5.1 States of matter

The arrangement and energy of particles determines whether a substance is a solid, a liquid or a gas.

Matter in the world around us takes various forms, and physics helps us to understand these in terms of their inner structures. Different laws of physics can be used to describe the behaviour of different states.

We're most familiar with matter in the form of solids, liquids and gases. Solids occupy a fixed volume and generally retain their shape without the need for a container. Their atoms are linked together by chemical bonds to form larger units called molecules. Molecules in turn bond together to build up repeating geometric structures as seen in many metals and minerals.

Supply sufficient heat energy to a solid, and the bonds between its atoms or molecules are weakened, allowing them to move past each other and rearrange themselves. Now a liquid, the matter occupies a fixed volume but changes shape to fit the environment.

When even more energy is supplied, the bonds between the particles break and the matter becomes a gas. Any forces between the atoms or molecules are generally very weak. The behaviour of gases is described in Topic 1.9.

A fourth state of matter is plasma. This is a fluid of even higher energy, composed of atoms that have lost their outer electrons and are therefore no longer electrically neutral.

Neon signs rely on plasma to carry an electric current and light them up.

Changing states

A plasma consists of charged ions and electrons that fill available space

Gas particles have no bonds with each other and therefore expand to fill the space available

Liquid particles retain a set volume, but can flow to change shape. A liquid settles at the bottom of a container

MORE ENERGY

A solid retains its shape regardless of any container

- Atoms or molecules
- Electrons
- Charged ions

Adding heat to a substance is the most common way of getting it to change its state. Other changes to the environment – for example, lowering or increasing pressure – can also trigger some changes of state.

5.2 Chemical elements

If atoms are the building blocks of matter, then the elements are the varieties of blocks available.

All atoms have distinct physical properties that govern their interactions. During the 19th century, the realization that some elements share such properties led to development of what we refer to today as the periodic table (see opposite).

Within the table, each element has a unique atomic number. Always a whole number, this denotes the number of protons in the element's central nucleus (see Topics 5.3 and 5.4). Individual atoms also have a mass number – that is, the sum of the protons and neutrons in the nucleus.

Few elements exist naturally in their pure form; most are chemical compounds, bonded together with atoms of other elements. The speed and ease with which an element binds to others is known as its reactivity. This relates directly to the structure of its atoms (see Topic 5.4). The most reactive element of all is fluorine. It is so chemically volatile that, in nature, it can only be found bound up with other elements.

Element 117 is so unstable it lasts for only a fraction of a second before decaying.

So far, 118 unique elements have been discovered. It is thought that 94 of these exist in nature – the others have to be created artificially. The balance of protons and neutrons in heavier elements can make them unstable. This means that they are prone to disintegrating to create other atoms through the process known as radioactive decay (see Topic 7.1).

The periodic table

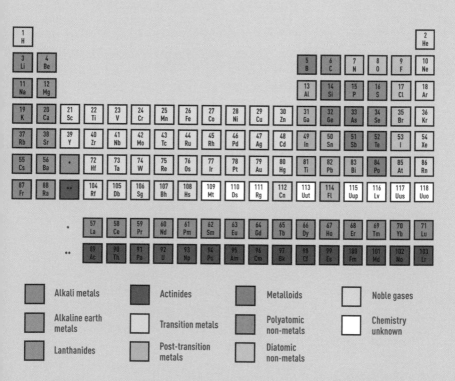

The periodic table is a way of grouping elements together in terms of shared characteristics – elements in the same column or group have similar chemical properties. They have similar electron configurations, albeit at different energy levels.

5.3 Electrons, protons and neutrons

Atoms are made up of three types of particle – protons and neutrons in a central nucleus, with electrons in orbit around them.

The discovery that atoms are composed of smaller particles provided a whole new area for scientists to explore – the subatomic realm. Initially, scientists thought atoms were fundamental, indivisible units of matter. But from the late 19th century, experiments gradually revealed more about their internal structure.

In 1897, English physicist J. J. Thomson showed that the cathode rays generated in electrical discharge tubes could be bent by a magnetic field. He concluded that the rays must be made up of unknown particles with a negative charge, now called **electrons**. He incorrectly thought that the electrons were embedded inside an atom, like fruit in a cake.

Thomson's one-time student, Ernest Rutherford, later proved him wrong by firing positively charged alpha particles at gold foil. He was expecting the particles to go straight through, but in fact some scattered in different directions. This could only be explained if most of the atom's mass was concentrated in a small space – the nucleus – with electrons orbiting around it.

Rutherford went on to find that many elements could be made to release positively charged particles when bombarded with radioactivity. Convinced he had found a fundamental building block of the nucleus, he named it the **proton**. In 1932, James Chadwick confirmed the existence of a third subatomic particle in the nucleus – the electrically uncharged **neutron**.

A proton has 99.8 per cent of the mass of a neutron.

Gold foil experiment

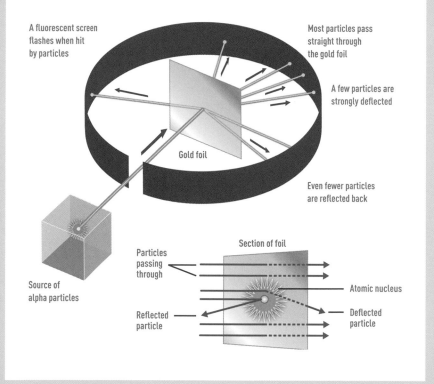

A fluorescent screen flashes when hit by particles

Most particles pass straight through the gold foil

A few particles are strongly deflected

Gold foil

Even fewer particles are reflected back

Source of alpha particles

Section of foil

Particles passing through

Atomic nucleus

Reflected particle

Deflected particle

The famous gold foil experiment, often attributed to Ernest Rutherford, was in fact carried out by his assistants Hans Geiger and Ernest Marsden. It showed that most of an atom is just empty space, with mass concentrated in a central nucleus.

5.4 Atomic structure

Electrons are arranged in distinct rings with different energy levels around an atom's nucleus.

The discovery that atoms had an atomic nucleus with orbiting electrons didn't explain all their observed properties. Neither did it explain why some chemicals were highly reactive while others weren't. New theories provided some answers.

In 1913, a theory by Danish physicist Niels Bohr explained why hydrogen emitted radiation of very specific wavelengths when atoms of the element were excited. He suggested that the nucleus was surrounded by a series of orbits at fixed distances. The energy of hydrogen's lone electron depended on which of these orbits it occupied. When it dropped from a higher orbit to a lower one, the electron released its excess energy in the form of light with a particular wavelength.

The Bohr model did not explain the behaviour of atoms with more than one electron, but other scientists built on his ideas with **valence theory**. This proposed that all electrons were contained in orbital shells, each of which can hold a certain number of electrons. The most stable configuration for an atom is to have a complete outer shell, and chemical reactions involve atoms gaining or losing electrons to reach this state.

Though generally accepted, valence theory still didn't explain everything about atoms. That would take entirely new physics – quantum theory (see Chapter 8).

Wavelengths of light emitted by excited atoms form a fingerprint for each element.

Simple atomic structures

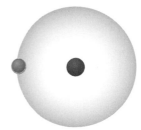

Hydrogen is the simplest element, with just a single proton in its nucleus and one electron orbiting around it

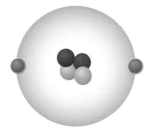

Helium, the second simplest element, has two protons and two neutrons in its nucleus, and two orbiting electrons

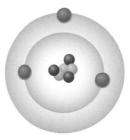

Lithium has three protons and three neutrons, and three orbiting electrons, one of which occupies a second energy level. This makes it highly reactive

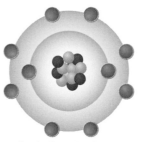

Neon has ten protons, ten neutrons and ten electrons filling both of its innermost shells. This stable configuration makes it chemically inert

Schematics of some simple elements shows how electrons fill up energy levels at increasing distances from the nucleus. These different energy levels are often called shells. Spare electrons and gaps in the outermost shell control an element's chemical reactivity.

5.5 Ionisation

When electrons are added to, or removed from, atoms they become electrically charged ions.

Individual atoms have balanced numbers of protons and electrons, making them electrically neutral. Yet if they lose or gain electrons they also gain a charge. This explains phenomena such as electricity.

When atoms become positively or negatively charged, they are known as **ions**. Elements are most susceptible to **ionisation** if they have a couple of electrons in an otherwise empty outer shell, or a near-complete shell with just one or two gaps (see Topic 5.4). One of two things can happen: an element acquires extra electrons from its surroundings, taking on an overall negative charge; or the element sheds electrons to other atoms to become positive overall.

Ionisation can happen in various ways. It may occur as part of a chemical reaction, leading either to an ionic bond between two oppositely charged particles (see Topic 5.7), or to free-floating ions in a solution. Acids, for example, separate into their constituent ions when dissolved. Bombarding some substances with electrons can create negative ions. Alternatively, strong electromagnetic fields can strip away weakly bound electrons, forming positive ions.

Fluorescent lightbulbs ionise mercury vapour inside them to produce light.

When light of specific frequencies falls on some materials, it causes electrons to escape and create ions. Known as the photoelectric effect, this was one of the discoveries key to understanding the true nature of light.

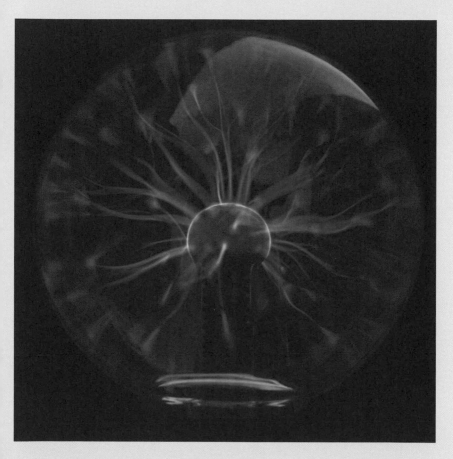

The shifting patterns of light in a plasma globe result from a phenomenon called coronal discharge: gas molecules break down into ions around a high-voltage electrode at the centre, and light is emitted by electrons switching between energy levels.

5.6 Emission and absorption spectra

Atoms have the ability to emit or absorb light at definite wavelengths.

The configuration of electrons within shells is unique to each element, but electrons can move between shells. When they do, they produce a spectrum – this is a series of lines that provides a fingerprint that can be used to identify an element.

Supplying energy in the form of electromagnetic radiation can cause electrons to jump into outer shells with higher energies. They absorb the necessary energy from the radiation and enter an excited state. But such states are usually unstable and short-lived. The excited electron soon drops back to the lowest available shell, shedding its excess energy in the form of emitted light.

The slight difference in the energy levels of different shells from one element to another means that each event of this kind involves a specific and unique amount of energy. This means the light emitted has a characteristic frequency, wavelength and colour.

As well as being emitted, light can also be absorbed. When light from a star passes through a gas, specific wavelengths are absorbed, revealing themselves as dark lines when the light is split into a spectrum.

This is the basis of spectroscopy – a powerful technique for analysing the composition of materials.

Absorption spectra help astronomers determine the temperature of stars (see Topic 4.8).

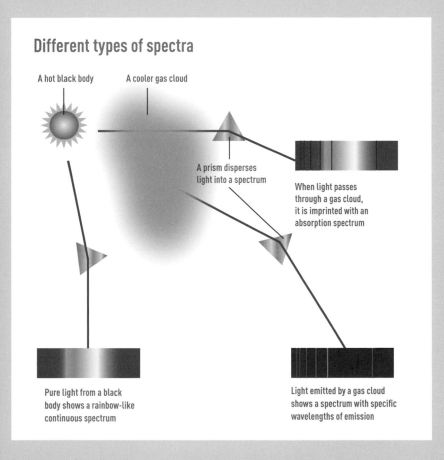

Different types of spectra

A hot black body

A cooler gas cloud

A prism disperses
light into a spectrum

When light passes
through a gas cloud,
it is imprinted with an
absorption spectrum

Pure light from a black
body shows a rainbow-like
continuous spectrum

Light emitted by a gas cloud
shows a spectrum with specific
wavelengths of emission

Emission and absorption spectra are useful in astronomy because light is our only means of studying faraway objects. They are also a powerful tool for analysing atomic structure in the laboratory.

5.7 Chemical bonds

Atoms bond in various ways to give materials different physical properties and chemical reactivities.

The most stable configuration for any atom is to have a complete outer shell of electrons. Atoms transfer electrons from one to another in order to reach this state. The result is the formation of bonds between them, and this is the basis of many chemical reactions.

■ Ionic bonds (see Topic 5.5) involve the formation of electrically charged ions by the addition or removal of one or more electrons from the outermost shell. When electrons are added, they create an **anion** with an overall negative charge. When they are removed, they result in a positive **cation**. Ionic bonds usually form when one type of atom, with a nearly empty outer shell, sheds electrons. Another type, with a nearly full shell, gains them and **electrostatic forces** bind them together (see Topic 5.9).

■ Covalent bonds involve a sort of timeshare between atoms that both want to add electrons to their outer shells. Each atom donates an electron to a shared pair that belongs partially to both. This results in a very strong bond.

The chemical bonds between hydrogen and oxygen in water are covalent.

■ Metallic bonds occur between large numbers of atoms that need to shed excess electrons in order to achieve stability. These electrons form a shared sea of negative charge that binds them together in a lattice-like structure. The **delocalized** electrons explain some metallic properties, such as malleability, lustre and ability to conduct electricity.

Different kinds of bonding

IONIC BONDING

A lone electron from one atom's
outer shell is donated to complete
a shell in the other atom

Electromagnetic force bonds
positive catirons and negatively
charged anions

COVALENT BONDING

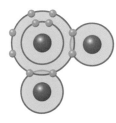

Atoms that each need to acquire
electrons to complete their outer
shell do so by creating shared pairs

METALLIC BONDING

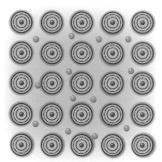

Atoms with excess electrons (shown in blue)
shed them to create a lattice of positive ions
amid a sea of electrons

Three types of chemical bond are formed when different types of element react together.
Non-metals (with near-complete electron shells) can form either ionic or covalent bonds.
Metallic elements tend to bond by giving up electrons in either metallic or ionic bonding.

5.8 Chemical reactions

When chemicals react together, old bonds break and new ones form. Electrons are redistributed as a result.

When two or more chemical compounds are brought into contact, they often undergo a chemical reaction that involves an exchange of electrons and a recombination of atoms into different compounds.

A substance losing electrons is said to be **oxidized**, while one that gains them is (somewhat confusingly) **reduced**. Although oxidation doesn't have to involve oxygen, a common example of a reaction in which oxygen is present is the rusting of iron, which results in a red compound called iron oxide.

Chemical reactions are often written down as equations. For example, the combustion of methane and oxygen (the reactants) produces carbon dioxide and water (the products). It can be written as:

$$CH_4 + 2O_2 \rightarrow CO_2 + 2H_2O$$

Significantly, the total number of atoms on each side of the equation is always the same, or balanced. The simple ratios in which substances react with each other, and the products formed as a result, were once an important clue in identifying many of the elements.

Reactions are either **endothermic** or **exothermic** – they either absorb energy or release it. Firework displays are spectacular examples of exothermic reactions.

Baking involves an endothermic reaction when heat from the oven is absorbed.

Thermite reactions are violent reactions that take place when a powdered metal such as aluminium is mixed with an oxide such as iron oxide, and ignited. The aluminium steals oxygen from the oxide, releasing huge amounts of heat in the process.

5.9 Van der Waals and other forces

Surprisingly strong attractive or repulsive forces can result from an uneven distribution of electric charge within molecules.

So far we have discussed strong chemical bonds. There are other types of bonds, however, that bind atoms and molecules together more weakly. Collectively known as intermolecular bonds, these are mostly electrostatic (see Topic 5.7). They are responsible for many natural phenomena.

Van der Waals forces are the most widespread example of electrostatic bond. They usually arise because the electrons in a covalent bond (see Topic 5.7) are constrained in their movement, creating a molecule whose negative charge is not evenly distributed. The concentration of electrons in one area gives it a slight net positive charge elsewhere – a so-called electric dipole whose charged ends can be attracted to the oppositely charged regions of similar molecules, or even to more strongly charged ions. In nature, the van der Waals force enables a gecko to walk up walls and hang from ceilings (see opposite).

A stronger type of intermolecular bonding occurs when positively charged hydrogen nuclei within molecules are attracted to negative charge centres in other molecules. This hydrogen bonding is seen in a number of compounds, but is particularly strong in water where the hydrogen bonds to the electronegative oxygen atoms of its neighbours.

Hydrogen bonding has significant effects. It raises water's melting and boiling points and gives it a high surface tension.

Water striders are able to walk on the surface of a pond because hydrogen bonding holds the water's surface molecules together.

Gecko feet have hair-like structures called setae that create temporary dipoles in their molecules that allow them to stick to almost any surface through Van der Waals forces. Human engineers have developed several technologies that aim to mimic this natural ability.

5.10 Mass spectrometry

Vaporizing a substance and turning it into ions is one way of revealing its inner substance.

How can we tell exactly what atoms, isotopes and molecules are present in a substance? Spectroscopy offers one effective method (see Topic 5.6). The most direct way of detecting small traces and measuring their exact proportions, however, is via **mass spectrometry**.

The technique involves breaking a sample of material down into component particles, turning them into ions, separating them by mass and measuring the abundance of each kind of ion in the sample.

Creating a vaporized cloud of electrically charged particles (ions) is often done by bombarding a sample with high-energy electrons. The ions are accelerated by an electric field, escaping in a narrow beam that passes into a mass analyser. This uses a magnetic field to deflect ions from their original path – lighter ions with the same charge are deflected more than heavier ones. Finally, a detector records the spot where each ion hits. In this way, the relative abundance of each ion is measured, allowing the number of different isotopes in the sample to be calculated.

A mass spectrometer was taken to Saturn's moon Titan to measure the composition of its atmosphere.

Mass spectrometry is used for applications such as radiometric dating (see Topic 7.8), detecting the chemicals present on other planets, and in forensic science to reveal the presence of poisons and toxins in very small amounts.

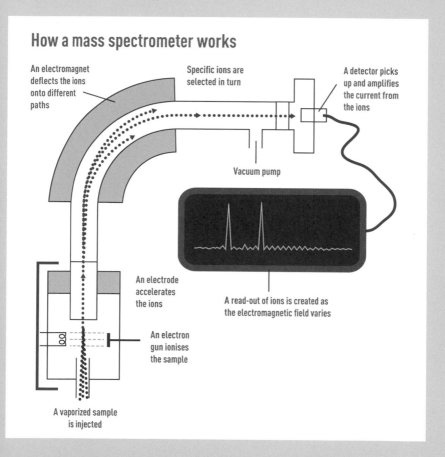

How a mass spectrometer works

An electromagnet deflects the ions onto different paths

Specific ions are selected in turn

A detector picks up and amplifies the current from the ions

Vacuum pump

An electrode accelerates the ions

A read-out of ions is created as the electromagnetic field varies

An electron gun ionises the sample

A vaporized sample is injected

A mass spectrometer measures only the current from ions with a specific mass-charge ratio at any one time. By varying the strength of the electromagnet, different ions can be brought into focus, building up a complete profile of the sample.

ELECTRICITY
AND MAGNETISM

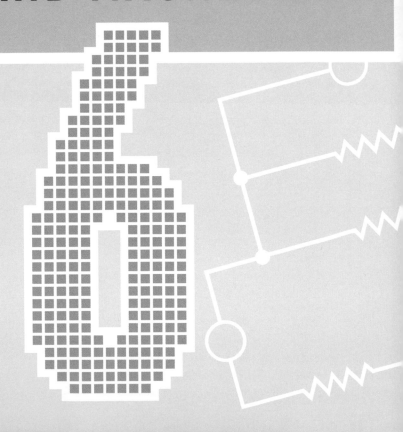

The physical phenomena of electricity and magnetism lie at the heart of the modern world – we rely on them for everything from domestic water heating to entertainment. It is difficult to imagine a time when electricity was not ubiquitous.

Yet for much of human history, both phenomena were curiosities to be marvelled at. Early man discovered that rubbing fur on amber created sparks. Today we understand that the sparks are bursts of static electricity. The rubbing action transfers electric charge from one substance to the other. Opposite charges are created as a result, which then attract one another. More commonly electricity was only encountered in lightning storms, while magnetic phenomena were restricted to naturally magnetic iron ore, known as the lodestone, which ultimately gave rise to the first magnetic compasses.

Continues overleaf

English physician William Gilbert published an investigation into magnetism in 1600, yet progress did not accelerate until around 1800. It was then that Italian physicist Alessandro Volta found that certain chemical reactions produce electricity, and subsequently used this phenomenon to build the first primitive batteries. It was this source of readily available electricity, above all, that stimulated the huge boom in exploration and discovery during the 19th century.

Many of the principles discovered during that time underlie today's modern electronic society. This chapter dedicates some considerable space to them, as well as to the atomic and subatomic causes of static and current electricity. Towards the end of the chapter we turn our attention to the underpinnings of our modern technological society – the solid-state semiconductors that have worked their way into every aspect of our lives, and the binary logic behind their design.

Contents

6.1 Static and current electricity

The effects of electricity are all around us, from spectacular displays of lightning to the power supplies in our homes.

Electricity is the result of electric charge, which creates **static electricity** when stationary and **electric current** when in motion.

Electric charge is due to small, negatively charged particles inside atoms – electrons. When atoms have the same number of electrons as positively charged protons, they are neutral. When they lose electrons, they become positively charged, and when they receive extra electrons they become negatively charged. (See also Topic 5.5).

An example of static electricity can be seen in clouds when ice particles collide. Some particles lose charge while others gain it. When the two charges become separated and build up, an electric current is eventually discharged between them – lightning. Charge is measured in coulombs (C), with 1 C equivalent to the charge on 6.24×10^{18} protons or electrons.

Current involves the flow of electrons through a conducting material. Metals are good conductors, because their electrons are only loosely bound to atoms and can move around freely. In most conductors, interaction between the moving electrons and the surrounding material creates resistance that reduces the flow of current. So-called **superconductors**, however, have no resistance at all thanks to the strange effects of quantum physics (see Topic 8.10).

Astronauts have observed enormous electrical discharges, called sprites, that escape upwards above certain types of thunderstorm.

A lightning strike is a spectacular form of electric discharge, created when movement in storm clouds leaves an excess of negative charge at their base. This eventually finds its way to the ground by breaking down and ionizing the air in between.

6.2 Voltage, cells and circuits

Voltage is the force that pushes charge-carriers through conducting materials.

The physics underlying current helps engineers design new electrical devices and make existing ones more efficient.

Electric current may move steadily in one direction (direct current; DC) or continuously change its direction (alternating current; AC, see Topic 6.7). It flows as the result of an electric potential or voltage between two points. This is a measure of the difference in potential energy between the two points in question based on their positions in an electric field. Think of the potential energy that a boulder has at the top of a mountain, owing to gravity, compared to one that's rolled to the bottom. The potential difference between two points is measured in volts and is defined as the energy required to move 1 coulomb of charge between them. The current flows from high to low voltage.

Devices that create a voltage include batteries. These typically have two connections or **electrodes** with a voltage difference between them. According to the conventions of a circuit diagram (see opposite), the direct current flows from the positive electrode or **anode** towards the negative electrode or **cathode**. It passes along connected wires and components that resist electrons attempting to move through them. Resistance dissipates some energy in the form of heat, which is why gadgets often feel hot.

Dividing the voltage across an electrical component by the current flowing through it gives its resistance.

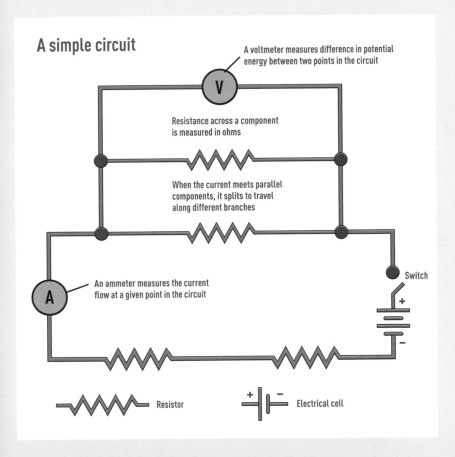

A simple circuit

A voltmeter measures difference in potential energy between two points in the circuit

Resistance across a component is measured in ohms

When the current meets parallel components, it splits to travel along different branches

An ammeter measures the current flow at a given point in the circuit

Switch

Resistor

Electrical cell

Engineers and physicists use symbols to illustrate circuit components. Conventional current is treated as emerging from the positive terminal of the electrical cell and returning to the negative terminal – in fact, electrons flow the opposite way.

6.3 Magnetic attraction

Magnetism results when electric charges move around. It happens inside atoms as well as through wires.

Magnetic materials are created when hidden electric charges inside atoms are influenced by a magnetic field. But only certain materials have the right properties to become magnets.

Inside atoms, negatively charged electrons orbit the nucleus. They also have intrinsic angular momentum, or spin (see Topic 8.8). Together, these give electrons a **magnetic moment**. They act just like normal bar magnets with north and south poles, where opposite poles attract and like poles repel. In most atoms, electrons pair up and their magnetic fields cancel each other out. But some materials have atoms with unpaired electrons. If enough of their magnetic moments line up in the same direction, the material as a whole is magnetic.

Certain materials, such as iron and nickel, are **ferromagnetic**. That is, they have some regions – called domains – in which all the magnetic moments line up, and other regions in which they do not. Exposure to a magnetic field enlarges the size of a lined-up domain to make a material noticeably magnetic. Even if a material does not show noticeable magnetism, the presence of magnetic moments inside it determines whether it is influenced by other magnetic fields.

The tendency of magnetic materials to line up with Earth's magnetic field was used for navigation in China and Europe from the 12th century.

Magnetism has many applications that make use of the way magnetic fields interact with electricity. It plays a key role in the electric motor and generator, for example, and also in the electrical transformer and electromagnet (Topics 6.5 and 6.6).

Magnetic field lines

Field around an isolated magnet

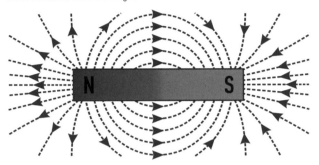

The field conventionally points away from the north pole

Field between opposite poles

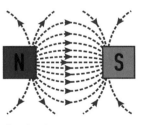

Opposite poles attract

Field between similar poles

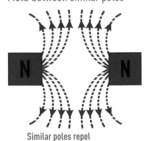

Similar poles repel

Magnetic field lines help to demonstrate magnetic fields. The direction of the lines indicates the way a magnetic needle would deflect if placed at each location. The density of the lines indicates the strength of the field in different places – for example, nearest the poles.

6.4 Forces, fields and induction

A moving magnetic field can be used to make electric currents flow.

When electric charges are close to one another, they experience a force, either of repulsion or attraction, depending on whether the charges are similar or opposite.

The force is given by Coulomb's law. Like Newton's law of gravitation this is an inverse square law. It states that the strength of the force is proportional to the product of the two charges and inversely proportional to the square of the distance between them. A similar force is created between two parallel wires carrying a flowing current and is defined by a fairly similar equation known as Ampere's law.

Both of these effects arise because a moving charge creates a magnetic field around it, influencing the magnetic moments in other materials (see Topic 6.3). A changing magnetic field also influences charge-carriers in conducting materials. It creates an electromotive force (emf) that causes a current to flow. This effect, known as electromagnetic induction, was discovered by Michael Faraday around 1831. His law of induction states that, for an idealized conducting wire, the induced emf is proportional to the rate of change in the magnetic field passing through it. Induction is key to most forms of large-scale electricity generation (see Topic 6.5).

In practice, induction is complicated by the fact that induced electrical currents create magnetic fields of their own.

In 1873, James Clerk Maxwell used Faraday's results to form a mathematical framework uniting electric and magnetic fields in a wave theory of electromagnetism.

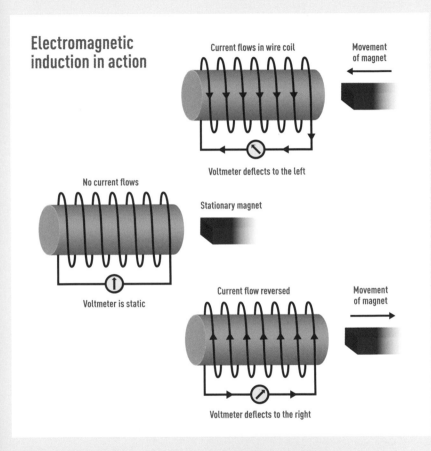

Electromagnetic induction in action

Current flows in wire coil

Movement of magnet

Voltmeter deflects to the left

No current flows

Stationary magnet

Voltmeter is static

Current flow reversed

Movement of magnet

Voltmeter deflects to the right

When a magnet is moved towards a wire coil, the changing magnetic field induces an electromotive force (detected by the voltmeter), and current flows. When the magnet is withdrawn, both force and direction of the current reverse.

6.5 Motors and generators

Many mechanical processes harness electric current in order to generate physical work.

From electric toothbrushes to food mixers, all electrical appliances rely on flow of current. A motor uses changing currents to create a constant force and turn a rotor, while a generator uses induction from a moving magnetic field.

In a motor, changing the flow of current in a conducting loop (the rotor) generates a magnetic field that deflects the rotor away from another fixed magnetic element (the stator). As the rotor's field deflects it away from the like poles of the stator, it becomes aligned with the attractive, unlike poles. To prevent it coming to a halt, a device called a commutator reverses the current with each half-rotation so that the rotor is always deflected.

Electric generators – from bicycle dynamos to steam turbines – effectively reverse this principle to make use of electromagnetic induction (see Topic 6.4). There are many approaches, but the simplest use external mechanical force to drive the rotation of a drum wrapped with many loops of wire in an enclosure lined with magnets. The constantly changing magnetic field experienced by the wires generates an electromotive force (voltage) and causes current to flow.

Generators are used to save energy in the regenerative braking found in electric cars. When the brakes are on, the energy supplied by the electric motor is diverted to a generator that creates electricity to top up the battery.

The first true electric motor was built by the German Moritz Jacobi in 1834.

A simple electric motor

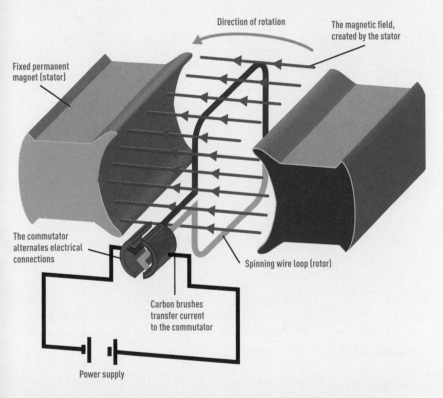

Direction of rotation

The magnetic field, created by the stator

Fixed permanent magnet (stator)

The commutator alternates electrical connections

Spinning wire loop (rotor)

Carbon brushes transfer current to the commutator

Power supply

A DC motor, like the model shown here, consists of a wire loop located between two magnets of opposite polarities (the stator). Current flowing into the wire causes it to rotate, and the motion is sustained by a commutator that repeatedly changes the direction of the current.

6.6 Electromagnets and transformers

Wrapping conducting wires around magnetic materials amplifies their magnetic fields.

While some materials are naturally magnetic, much more powerful magnets can be made using electric current. This principle is put to work in a wide range of technology.

Moving charge in a conductor creates a magnetic field. This is harnessed by **electromagnets**, the simplest of which is a long wire looped around a cylindrical core. When electricity flows through the wire, the arrangement mimics the field of a bar magnet, with poles at either end of the core (see Topic 6.3).

If the core is made of a ferromagnetic material such as iron, the effect is hugely magnified, as magnetic moments within the core's atoms line up with the field created by the current. Such electromagnets are stronger than permanent magnets and their strength can easily be controlled. They're used in everything from loudspeakers to particle accelerators.

Transformers, meanwhile, use looped conducting wires and a ferromagnetic core alongside electromagnetic induction to manipulate current and voltage. A varying current in a primary winding on one side of a square transformer core induces a magnetic field within the core, which in turn causes current to flow through a secondary winding. By varying the number of coils in each winding, voltages can be stepped down to a safe level. Appliances like laptops and mobile phones have transformers built into their power plugs for this purpose.

The electromagnet was invented in 1824 by British scientist William Sturgeon.

Step-down transformer

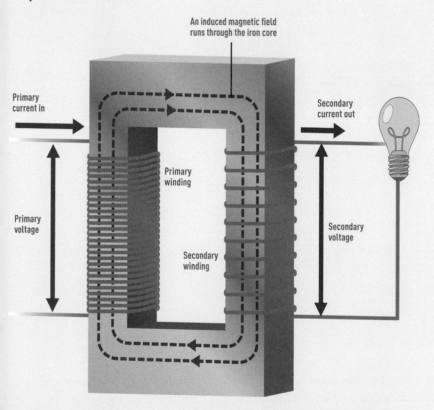

An induced magnetic field runs through the iron core

Primary current in

Secondary current out

Primary winding

Primary voltage

Secondary winding

Secondary voltage

Current flowing around the primary winding of a transformer creates a magnetic field in the core. This, in turn, produces current in the secondary winding through electromagnetic induction. The effect only works when the primary current is alternating current (Topic 6.7).

6.7 Alternating current

Creating a current that oscillates back and forth in its conductor overcomes many of the practical difficulties in electricity distribution.

The amount of current a conducting material can carry is limited by the number of charge-carrying electrons that can flow through it. Large currents require physically larger wires. The stronger the current and the further it has to flow, the more energy it loses to resistance. As a result, sending direct currents over long distances presented huge challenges for pioneers of electrical power.

Most electrical devices simply require charge-carriers to be in motion, regardless of their actual direction and so rely on direct current (DC) transmission (see Topic 6.2). In the 1880s, various engineers realized that they could transmit electricity as **alternating current** (AC) – a flow that oscillates or changes direction many times per second, with the charge-carriers themselves only moving small distances through a conductor.

This means that very high voltages can be used to push small currents through power cables over long distances without losing energy. Transformers can then lower the voltage and raise the current to do useful electrical work. AC transformer and motor designs are simpler than DC (since the induction effects they harness require a constantly changing voltage) while many other devices work equally well with either. Most electronic devices – TVs, computers and anything with a battery – use DC. For these, AC is transformed into DC using a rectifier circuit.

The fierce battle between rival engineers backing AC and DC electricity around 1890 is known as the War of the Currents.

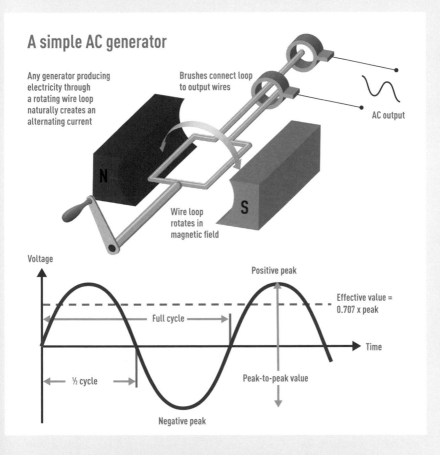

A simple AC generator

Any generator producing electricity through a rotating wire loop naturally creates an alternating current

Brushes connect loop to output wires

AC output

N

S

Wire loop rotates in magnetic field

Voltage

Positive peak

Effective value = 0.707 x peak

Full cycle

Time

½ cycle

Peak-to-peak value

Negative peak

The graph above shows how voltage varies over time, and the so-called effective value of the voltage delivered. The effective value is the level of DC current that would be required to deliver the same power – around 230 volts in the UK.

6.8 Photomultipliers and CRTs

Vacuum tubes and electrodes are used to convert current in order to display images and to make highly sensitive detectors.

Currents aren't only useful for powering electrical appliances. They're also exploited in devices that can detect low-intensity light or generate an image. Both were hugely important in the early days of television.

Photomultipliers detect low amounts of light – right down to individual photons. They do this by multiplying the current produced when photons hit a **photocathode** – a detector that emits electrons by the photoelectric effect (see Topic 5.5). The electrons pass through a vacuum tube, where they are attracted to a positively charged **dynode**. Electrons released by one dynode pass to the next, with more electrons released from each dynode. This cascade effect produces a stream that finally reaches the **anode** at the other end of the tube, where it is transformed into a current. The technology is used to amplify weak signals in night-vision cameras, as well as to detect gamma rays in medicine and astronomy.

TVs and computer displays were originally **cathode-ray tubes** (CRTs), which turn current into an image. This is a flask-shaped vacuum chamber, with a cathode at the narrow end and anodes along the neck. The current that enters is used to heat the cathode, creating a beam of electrons (cathode rays). As it passes between the anodes, the beam direction is manipulated using the anodes, before striking a phosphorescent screen to build up a picture.

The very first television system, built in 1911, used radio waves to send signals to a cathode-ray tube.

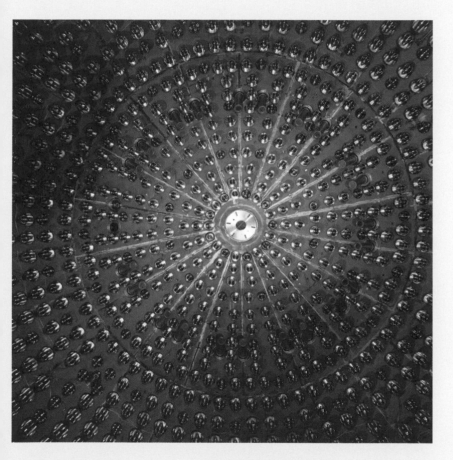

The Borexino experiment is a huge detector for elusive neutrino particles (see Topic 7.10), situated beneath a mountain at Gran Sasso in Italy. At its heart is this huge chamber lined with highly sensitive photomultiplier tubes.

6.9 Semiconductors

Some materials have both conducting and insulating properties, making them ideal when it comes to controlling electric current.

Electrical **conductors**, such as copper wire, are extremely useful, as are **insulating materials** like plastic. But the modern world is only possible thanks to **semiconductors** – materials that have the properties of both.

Semiconductors are usually made from chemical compounds of elements in group 14 of the periodic table – typically silicon and germanium – in a regular, crystalline form. These compounds can be **doped** by adding impurities to gain either a natural excess of negatively charged electrons (an n-type semiconductor), or an excess number of holes that act as though they have net positive charge (a p-type semiconductor). Addition of these impurities to a crystal can be used to create neighbouring p-type and n-type regions on the same crystal. Semiconductors can therefore permit both electron conduction (the actual movement of electrons through the crystal lattice), and hole conduction (the effective movement of positive holes in the opposite direction as electrons vacate their previous positions).

Semiconductors are used for manufacturing electronic components such as diodes, which permit current flow in just a single direction. These were the basis of the first **logic gates**, which perform logical operations to produce an output. Logic gates are at the heart of all computers (see Topic 6.10).

Doping allows semiconductor chips to be printed with micro-electronic components.

How a diode works

A semiconductor diode consists of p-type and n-type regions separated by a depletion region lacking charge carriers

In a reverse bias situation, a potential difference across the diode tends to draw charge carriers apart, so current cannot flow

In a forward bias situation, the potential difference across the diode is increased, allowing charge carriers to cross the depletion region so that current can flow

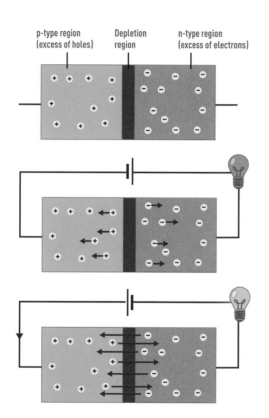

p-type region (excess of holes)   Depletion region   n-type region (excess of electrons)

Diodes are devices that allow current flow in one direction only. They are made by creating a p–n junction with an excess of positive holes on one side and an excess of electrons on the other. Current passes from the n-type side of the junction to the p-type side, but will not flow in reverse.

6.10 Digital electronics

Complex calculations involving huge amounts of data are made at the press of a button.

Electric current can normally take on any value, varying continuously between them in a wave-like or analogue way. Harnessing current to produce discrete digital values has made the modern world possible.

In computing, information is encoded as simple on/off pulses – ones and zeros – based on whether a signal is present or not. This **binary** system offers a powerful way of processing data. For example, two binary digits or **bits** can represent four possible states (00, 01, 10 and 11), eight bits (a **byte**) can represent 256 different values, and 32 bits can represent more than 4.2 billion values.

Streams of binary can represent anything from raw numbers to detailed instructions. Clever design allows the creation of systems of logic gates (see opposite) that can reconfigure themselves to carry out a huge variety of tasks. Combined with the development of semiconductor diodes and related devices (see Topic 6.9), the binary system opened the way for the digital age. Modern integrated circuits, usually etched on a single silicon or germanium chip, consist of countless tiny logic gates, each made from a combination of transistors and diodes, and able to perform logical comparisons between incoming binary signals.

One promising area of future research is optical computing, which harnesses photons to send data at the speed of light.

Digital electronics are designed using a form of algebra called Boolean logic (see opposite).

Logic gates

AND

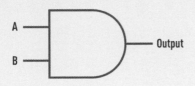

Inputs		Output
A	B	
0	0	0
0	1	0
1	0	0
1	1	1

An AND gate compares the currents at two input points A and B. Only if current is present at both, does it generate an output current

NOT

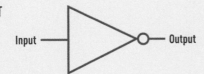

Input	Output
0	1
1	0

A NOT gate inverts the signal current it receives. It produces a current when there is no input, but does not produce a current when an input current is present

OR

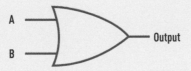

Inputs		Output
A	B	
0	0	0
0	1	1
1	0	1
1	1	1

An OR gate compares the current inputs at A and B. If there is a current at either or both of them, it generates an output signal

Logic gates are simple electronic components that take one or two input signals (a binary 0 or 1 depending on the level of current) and perform a simple logical task to deliver a binary output. In addition to the NOT, AND and OR gates shown here, there are several other types.

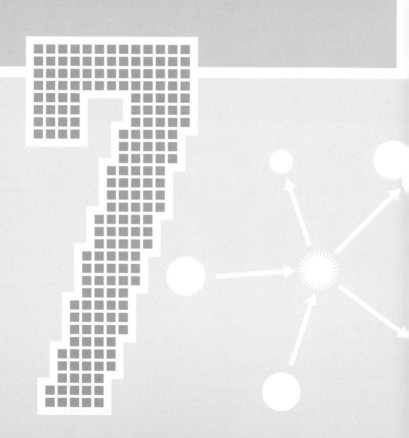

NUCLEAR PHYSICS

The atomic nucleus is a compact yet incredibly dense and energy-rich volume of space. Tiny even in comparison to an atom, it nevertheless contains almost all of the atom's mass in the form of subatomic proton and neutron particles. The idea of tapping into this submicroscopic realm might seem impossible, but with a helping hand from the laws of physics, scientists have been doing exactly this for more than a century.

That helping hand is natural radioactivity – the fact that certain configurations of protons and neutrons are inherently unstable, leading the nucleus to shed excess particles in an attempt to find a more stable state. The phenomenon was discovered in 1896 by French scientist Henri Becquerel when he found that a sulphate compound of uranium emitted a substance that caused photographic plates to fog. Marie Curie soon named the effect radioactivity, while New

Continues overleaf

Zealander Ernest Rutherford discovered that the emissions were in the form of two kinds of particles – a heavy, positively charged one and a much lighter, negatively charged one. Becquerel himself later added a third type of radioactivity to the list – the emission of gamma rays.

Radioactive substances offered a first look at the processes going on inside certain elements. Thanks to the lessons they revealed, later generations of scientists were able to explain how elements are formed in the first place, to understand why elements decay in certain ways, and even to discover the process that powers the Sun itself.

This chapter looks first at the essentials of nuclear physics, before discussing the ways in which physicists have learned to harness its power, and looking at some other unusual phenomena that have been discovered through radioactivity.

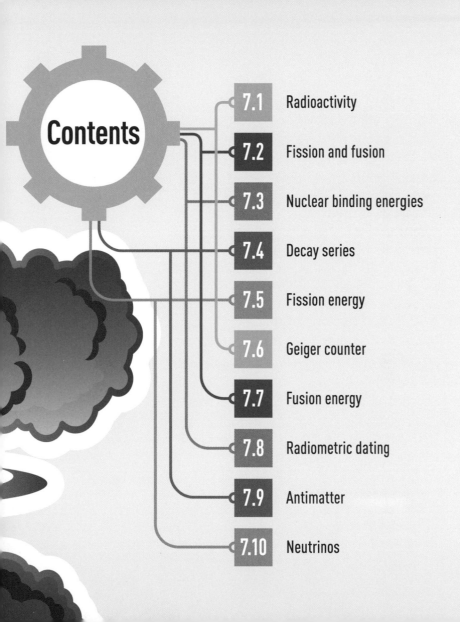

Contents

7.1 Radioactivity

Atoms sometimes need to shed particles or energy in order to become more stable.

The phenomenon of **radioactivity** was discovered in the 1890s by Henri Becquerel, and Marie and Pierre Curie. Working out the precise details took some time, but it became clear that it involves unstable isotopes of elements – that is, varieties of atoms in which an imbalance between protons and neutrons makes the nucleus unstable (see Topic 5.3). The nucleus undergoes radioactive decay in one of three main ways.

- **Alpha decay** involves the atomic nucleus shedding two protons and two neutrons, known as an alpha particle. This is effectively identical to the nucleus of the helium-4 isotope.

- In **Beta minus decay** a neutron in the nucleus transforms into a positively charged proton, and two particles are created and immediately ejected: a negatively charged electron and an antineutrino (see Topic 7.10).

- **Gamma radiation** involves the release, not of particles, but of excess energy in the form of high-frequency gamma rays.

Radioactive substances were sold in medicinal drinks before their dangers were discovered.

Since alpha particles have a large mass and charge, they travel only a few centimetres in air and can be stopped by a sheet of paper. Beta radiation travels further, but is stopped by a sheet of aluminum. It takes inches of lead or a large block of concrete to shield anything from gamma radiation. (For beta plus decay, see Topic 7.9).

Penetrating power

The dangers of different types of radiation
depend on a combination of their energy
and their penetrating power

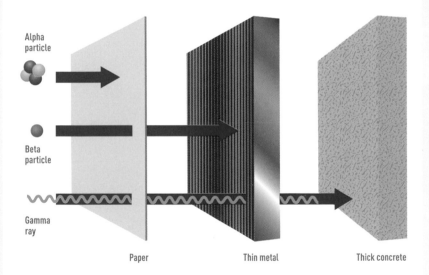

Alpha
particle

Beta
particle

Gamma
ray

Paper   Thin metal   Thick concrete

Gamma rays are generally regarded as the most dangerous type of radiation, because of their
penetrating power. However, alpha particles, although easily stopped if outside the body, can
be far more deadly than gamma rays if an alpha source is ingested.

7.2 Fission and fusion

Nuclear reactions come in two distinct varieties: one splits large isotopes into smaller ones; the other forms a new, larger nucleus from a number of smaller ones.

Nuclear reactions come in two distinct varieties, known as fission and fusion. Only fission occurs naturally on Earth. It is also the reaction harnessed in nuclear power plants.

■ In **fission**, a very large nucleus absorbs a neutron and splits into smaller fragments. One way of achieving this is to take an element with a large nucleus, such as uranium, and bombard it with slow-moving neutrons. The process also produces individual neutrons, which are used in nuclear power to create a chain reaction (see Topic 7.5). Radioactive alpha decay, shown opposite, is effectively a form of fission that occurs naturally and spontaneously, without the need for neutron bombardment.

■ In **fusion**, small nuclei join together to make larger ones. Unlike fission, fusion requires extreme conditions to take place – pressures more than 100 billion times that of Earth's atmosphere, and temperatures of 10 million °C (18 million °F) or higher. In such situations, the bare nuclei of elements collide with such energy that the repulsion between their positive charges is overcome and the short-range forces that bind nuclei together come into effect (see Topic 7.3).

Fusion may be harnessed to provide us with power in the future (see Topic 7.7) but it's already supplying us with energy. It's the power source of stars like the Sun, in which hydrogen nuclei are fused to make helium (see opposite).

Uranium-235 and plutonium-239 are isotopes commonly used for fission.

Nuclear reactions

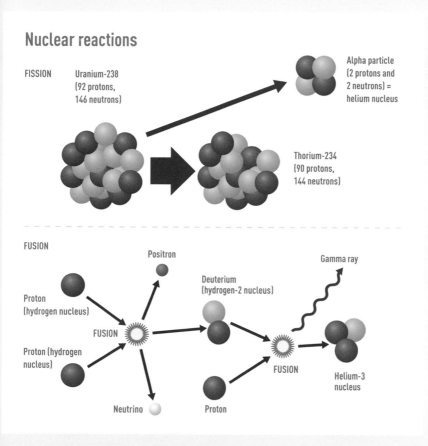

FISSION

Uranium-238
(92 protons,
146 neutrons)

Alpha particle
(2 protons and
2 neutrons) =
helium nucleus

Thorium-234
(90 protons,
144 neutrons)

FUSION

Positron

Proton
(hydrogen nucleus)

FUSION

Proton (hydrogen
nucleus)

Deuterium
(hydrogen-2 nucleus)

Gamma ray

FUSION

Helium-3
nucleus

Neutrino

Proton

These two diagrams depict common forms of nuclear fission and fusion. At top, the alpha-decay fission of the common uranium isotope U-238. Below, the initial stages of proton-proton fusion, the process that turns hydrogen to helium inside the Sun.

7.3 Nuclear binding energies

Why is it that some nuclei release energy when they split apart, while others do so when they join together?

Nuclei behave in different ways when binding, because the amount of energy locked up within an atomic nucleus varies from element to element. This **binding energy** of the nucleus explains why some nuclear reactions release energy, while others absorb it.

Just as energy is required to break chemical bonds and is liberated when new bonds form, the bonding of protons and neutrons in a nucleus also releases energy. And since energy and mass are equivalent according to Einstein's famous equation $E = mc^2$ (see Topic 10.5), the energy difference is reflected in the fact that the mass of the combined nucleus is very slightly less than the mass of its component parts.

When the binding energy of isotopes is compared to their mass, a pattern emerges, as follows:

- Fusing light elements to heavier ones releases energy.

- Attempts to fuse elements heavier than iron and nickel absorb energy.

- Fusion of hydrogen nuclei into helium releases more energy than the fusion of helium to create beryllium and carbon, as happens in stars older than our Sun.

- Fission of heavier elements such as uranium and plutonium also releases energy.

The isotope with the highest binding energy is nickel-62.

Balance of energies

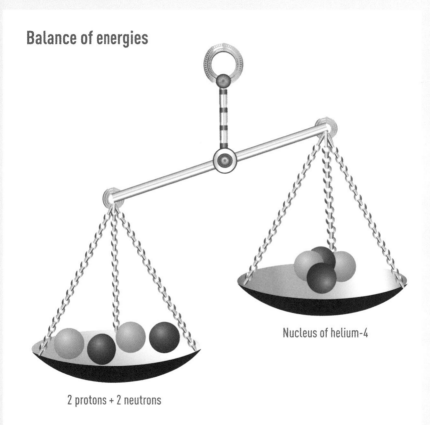

Nucleus of helium-4

2 protons + 2 neutrons

The surprising fact that a nucleus of helium (right) weighs less than its constituent protons and neutrons (left) is key to the way that stars like the Sun shine – for most of their lifetimes they fuse hydrogen into helium, transforming these small quantities of mass into large amounts of energy.

7.4 Decay series

Most radioactive materials undergo a complex series of decays before reaching stability.

When a heavy isotope spontaneously splits apart in radioactive decay, the daughter isotope created may be just as unstable as its parent. As a result, a single decay is usually just one step in a longer series, and the time it takes for a substance to decay into stable and harmless material determines how long nuclear waste remains radioactive.

Throughout any decay series, it is impossible to predict when an individual atom will decay. But it is possible to measure the rate at which decay is taking place and work out the isotope's statistical **half-life** – the time taken for half of a sample to decay from parent to daughter.

Depending on the structure of the nucleus, half-lives can vary from minuscule fractions of a second to billions of years. The isotope strontium-90, for example, undergoes beta decay with a half-life of 28 years, while the half-life of the nuclear fuel plutonium-239 is 24,000 years.

Since even a pure sample of a single radioactive isotope will eventually give rise to a complex mix of other isotopes, managing nuclear waste isn't easy. However, it is possible to separate residual uranium and plutonium from other decay products and reuse them as fuel in a nuclear reactor. This helps reduce the amount of waste.

Iodine-129 has a half-life of 15.7 million years.

Decay of uranium-238

This complex chain indicates the various pathways taken by the common uranium isotope U-238 as it decays towards stable lead-206. The most common paths are shown by black arrows, with the less likely alternatives in grey.

Alpha decay: reduces mass by 4 and atomic number by 2

Beta decay: transforms neutron to proton, increasing atomic number by 1

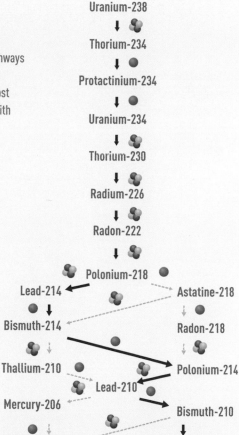

7.5 Fission energy

Fission reactors generate energy by sustaining a chain reaction, in which each decay event triggers at least one more.

Today's nuclear reactors harness the power of induced fission. Strike a radioactive isotope with neutrons, and it can be stimulated to decay earlier than it would naturally, so kick-starting a chain reaction. The energy released in this chain reaction is turned into electricity.

Nuclear reactors generate a chain reaction using fuel rods enriched with large amounts of a radioisotope, such as uranium-235, an isotope of uranium with a relative atomic mass of 235. Neutrons emitted during one decay event can trigger the next. In fact, the main challenge lies in slowing or moderating the chain reaction. If it isn't slowed, the results can be explosive – as happens in an atomic bomb.

About 80 per cent of the time, absorption of a neutron by uranium-235 triggers fission, producing two large daughter isotopes and three stray neutrons. It also releases binding energy (see Topic 7.3), which is extracted by using it to heat up a working fluid. For example, water is heated to boiling point to generate steam, which is used to drive electricity-producing turbines.

With a half-life of 8 days, iodine-131 is a highly radioactive product of fission.

Nuclear power plants don't release greenhouse gases, but they do generate toxic waste in the form of other radioisotopes that have far higher rates of radioactivity and shorter half-lives than uranium and therefore require careful handling.

Fission chain reaction

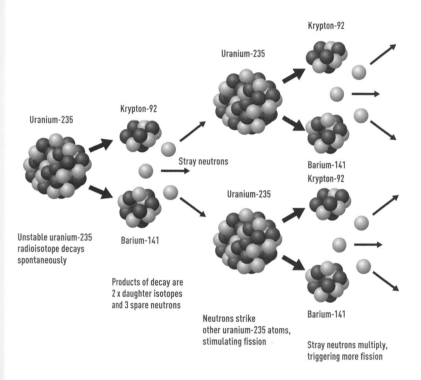

Uranium-235

Krypton-92

Uranium-235

Krypton-92

Stray neutrons

Barium-141

Uranium-235

Krypton-92

Unstable uranium-235
radioisotope decays
spontaneously

Barium-141

Products of decay are
2 x daughter isotopes
and 3 spare neutrons

Neutrons strike
other uranium-235 atoms,
stimulating fission

Barium-141

Stray neutrons multiply,
triggering more fission

Certain radioisotopes such as uranium-235 will undergo stimulated
decay when they are struck by a neutron with sufficient energy. If there
is enough radioactive fuel present, then a chain reaction can result.

7.6 Geiger counter

A Geiger counter detects radioactive emissions, such as alpha and beta particles or gamma rays.

The archetypal radiation detector is a device that emits audible clicks in the presence of ionising radiation in the form of alpha and beta particles and gamma rays (see Topic 7.1).

Developed in 1928 by Hans Geiger and Walther Müller, it consists of a handheld probe containing a Geiger–Müller tube. The tube is linked to a power supply and some form of read-out device, such as a meter or loudspeaker.

The tube has thick walls, usually with a thin window at one end, and is filled with a low-pressure gas. A wire through its centre acts as the anode in an electrical circuit, while the interior of the tube itself acts as the cathode. When an alpha or beta particle enters the detector, the gas briefly becomes conductive and an avalanche of electrons bridges the circuit, sending a pulse to the counter. Pulses sound the same for both alpha and beta radiation, so it's not possible to distinguish between them.

To detect gamma radiation, a windowless tube is used, since the rays pass straight through most materials. Detection relies on a small proportion of the rays striking a dense steel lining and releasing electrons, which then flow in the normal way.

Geiger counters measure clicks per minute (CPM) for alpha and beta radiation.

The essential principle of the Geiger counter is used in a wide variety of particle detectors, though solid-state electronic alternatives using semiconductor materials are increasingly common.

The Geiger-Müller tube

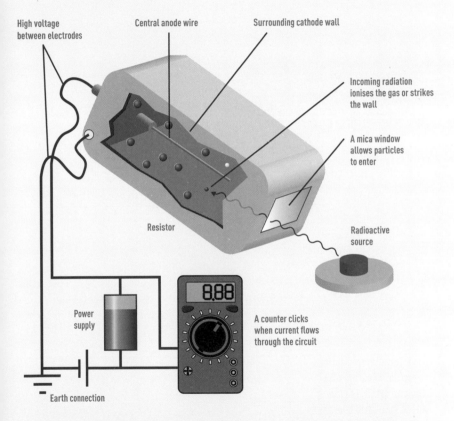

High voltage between electrodes

Central anode wire

Surrounding cathode wall

Incoming radiation ionises the gas or strikes the wall

A mica window allows particles to enter

Resistor

Radioactive source

Power supply

A counter clicks when current flows through the circuit

Earth connection

The instrument has a potential difference of several hundred volts between its central anode and surrounding cathode, but the circuit cannot conduct because its electrodes are separated by the gas. When a particle enters, it strikes a gas atom and ionises it, triggering a burst of electric current.

7.7 Fusion energy

How possible is it to recreate conditions normally found in the heart of stars?

In theory, fusion energy offers the potential for limitless, clean power. Just as nuclear fission releases binding energy (see Topic 7.3) when heavy nuclei split, so fusion releases energy when two light nuclei join together.

Reactions in the Sun fuse standard hydrogen nuclei (single protons) into helium nuclei, one step at a time (see Topic 7.2). Reactors on Earth attempt the more achievable final steps, fusing the heavy hydrogen isotopes deuterium and tritium to create helium and release energy. Deuterium (with one proton and one neutron in its nucleus) and tritium (with one proton and two neutrons) are rare in nature compared to normal hydrogen, but can be extracted from seawater.

Artificial fusion is still in the experimental phase. There are two main technologies, either using plasma contained in a magnetic field or lasers that blast a small pellet until fusion begins. Recently, **break-even** has been achieved, producing as much energy as was supplied in the form of fuel. To be practically useful, however, fusion would also need to produce more energy than is needed to run the reactor. Only at this point would a fusion reactor be truly self-sustaining. So far, then, the most impressive demonstrations of nuclear fusion seen on Earth have been in tests of hydrogen bombs such as Ivy Mike (see opposite).

The International Thermo-nuclear Experimental Reactor (ITER) in France will be the world's largest fusion experiment when it's switched on in 2027.

The first hydrogen bomb, code-named Ivy Mike and tested in 1952, used a small fission explosion to trigger a burst of nuclear fusion, releasing energy equivalent to 10.4 megatons of high explosive.

7.8 Radiometric dating

Radioactivity can be used to estimate the age of rocks and organic matter.

Among the best-known scientific applications of radioactivity is **radiometric dating**, where isotope ratios are used to calculate the age of materials. By measuring the amount of a radioactive substance with a known half-life (Topic 7.4) in a material, it's possible to estimate how old it is.

One of the most popular techniques is commonly known as carbon dating. It relies on the fact that all living things contain, not just carbon, but a small amount of the radioactive isotope carbon-14. While an organism remains alive, the internal ratio of its carbon isotopes remains in balance with the environment. But once it dies, the carbon-14 in its organic matter can no longer be exchanged and begins to decay at a predictable rate.

The ratio of radioactive carbon-14 to normal carbon-12 in once-living organisms allows scientists to calculate how many half-lives have passed since the material formed and this, therefore, indicates its age.

Because half-life is a statistical effect, it is never absolutely precise, and radiometric dates are best reported as possible date ranges.

The drawback with this method is that the isotope's relatively short half-life of 5,730 years makes measurements for older samples progressively less accurate as the carbon-14 dwindles away. The timescale is around 50,000 years. What's more, the future for carbon dating looks bleak, as human industrial activity is changing the ratios of carbon in the atmosphere.

Dating with carbon-14

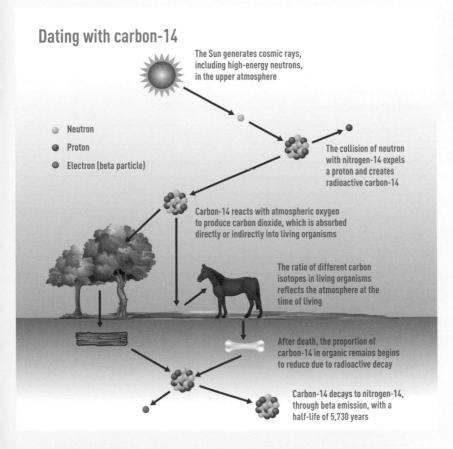

The Sun generates cosmic rays, including high-energy neutrons, in the upper atmosphere

- Neutron
- Proton
- Electron (beta particle)

The collision of neutron with nitrogen-14 expels a proton and creates radioactive carbon-14

Carbon-14 reacts with atmospheric oxygen to produce carbon dioxide, which is absorbed directly or indirectly into living organisms

The ratio of different carbon isotopes in living organisms reflects the atmosphere at the time of living

After death, the proportion of carbon-14 in organic remains begins to reduce due to radioactive decay

Carbon-14 decays to nitrogen-14, through beta emission, with a half-life of 5,730 years

Though the principle behind radiocarbon dating is simple, in practice these techniques must take into account all sorts of factors – even the fact that the additional weight of radioisotopes can cause them to concentrate in certain parts of the environment.

7.9 Antimatter

Antiparticles have an equal, but opposite, electric charge to their normal matter equivalents.

Antimatter is famous in fiction as the substance that powers *Star Trek*'s USS *Enterprise*. In physics, it is like the mirror image of normal matter, with the electric charges reversed.

Matter and antimatter particles annihilate on contact with each other, disappearing in a burst of gamma rays as their mass converts directly into energy in accordance with Einstein's equation $E = mc^2$. In theory, this could be harnessed as energy but in practice, it would be a huge challenge to store even small amounts without annihilation occurring.

The most common antimatter particle is the **positron**, a positively charged equivalent of the electron. It is produced by so-called **beta plus decay**, in which a proton in the nucleus of a radioisotope spontaneously changes into a neutron. To maintain the overall charge of the nucleus, a unit of positive charge is shed as a positron.

The fact that everything around us exists is an unsolved mystery. It's thought that matter and antimatter were created in the big bang in equal amounts, but if that were true everything would have been annihilated straight away.

1 kg (2.2 lb) of matter reacting with antimatter would produce energy equivalent to 43 megatons of TNT.

Antimatter is found in nature, in cosmic rays (high-energy particles from space), and can be made artificially. Applications include the medical imaging technique of PET (positron emission tomography).

Natural antimatter

BETA MINUS DECAY (ß⁻ DECAY)

In normal ß– decay, a neutron in the nucleus transforms into a positively charged proton. To keep the electrical charge balanced, an electron and an antineutrino are released

Carbon-14
(6 protons, 8 neutrons)

Nitrogen-14
(7 protons, 7 neutrons)

Antineutrino   Electron

BETA PLUS DECAY (ß⁺ DECAY)

In the rare ß+ process, a proton transforms into an uncharged neutron. Excess positive charge is shed by creation of a neutrino and a positron

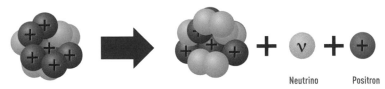

Carbon-14
(12 protons, 11 neutrons)

Sodium-23
(11 protons, 12 neutrons)

Neutrino   Positron

A few rare forms of radioactive decay naturally generate antimatter in the form of positrons. This happens when a radioisotope can reach a more stable configuration by transforming one of its protons into a neutron.

7.10 Neutrinos

Neutrinos are abundant, but elusive, particles and the ultimate messengers from outer space.

The elusive **neutrino** is a particle generated by some nuclear processes, including beta plus decay (see Topic 7.9). The fact that it has no electric charge and almost no mass makes it an ideal tool for studying the furthest reaches of the cosmos.

Produced in huge quantities by nuclear fusion in the Sun and other stars, and moving at close to the speed of light, countless neutrinos pass through our bodies all the time. In fact, most of the time they pass right through the entire planet since they rarely interact with everyday matter. Since neutrinos have no charge, they are not deflected by electromagnetic fields, and that means they reach us direct from distant stars. There are also antineutrinos, which share the same properties as neutrinos but have opposite spin (see Topic 8.8).

If neutrinos rarely interact, how are they detected? Neutrino observatories are typically buried far underground, where overhead rock blocks other particles. One detection method uses a huge tank of water (see opposite). Neutrinos very occasionally collide with other molecules, leading to the telltale blue flash of Cherenkov radiation (see page 105).

Neutrinos were once thought to be entirely massless, but experiments have shown that they do have mass – it's at least a million times smaller than an electron. And recently, it was discovered that neutrinos oscillate between three different varieties, only one of which can be detected at all.

Neutrinos are probably the second most common particle in the universe, after photons.

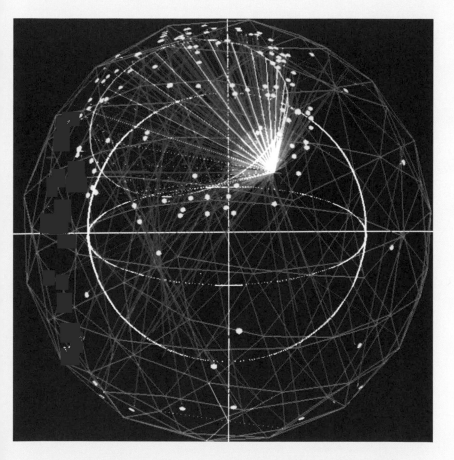

A computer simulation maps the track of a neutrino event in an underground water tank. On rare occasions when a neutrino strikes a water molecule head-on, a high-speed electron can be emitted, creating Cherenkov radiation that can be detected by a photomultiplier (see Topic 6.8).

QUANTUM
REALITY

Two revolutions in the early 20th century transformed physics forever. One was Einstein's theory of relativity (see chapter 10). The other was quantum physics – the discovery that, on the smallest scales, many of the normal certainties of our universe break down.

The quantum revolution began quietly enough at the turn of the century, with Max Planck's suggestion that light could sometimes be treated as if it consisted of discrete packets of energy rather than a continuous wave. Albert Einstein caused a stir in 1905 when he provided evidence that Planck's photons were a physical reality. But it was not until the 1920s that events gathered pace with Louis Victor de Broglie's suggestion that, not only can waves sometimes act like particles, but particles can sometimes act like waves.

Continues overleaf

Soon proved to be true, de Broglie's idea of wave-particle duality unleashed a storm of uncertainty on the previously serene world of classical physics. On a submicroscopic scale, it seems, particles really do act like waves, and this means they are not as predictable as we once thought they were.

This chapter can only hint at the extraordinary implications of quantum physics. There are scientists who have spent the best part of a century trying to reconcile this revolutionary idea with the predictable mechanics of the everyday world. Others are more concerned with the possible applications of quantum physics. Our whistle-stop tour aims to combine both approaches. We begin with a brief introduction to the strange world of quantum physics, then consider some popular interpretations of the theory, before finally turning to some of the astounding real-world consequences.

Contents

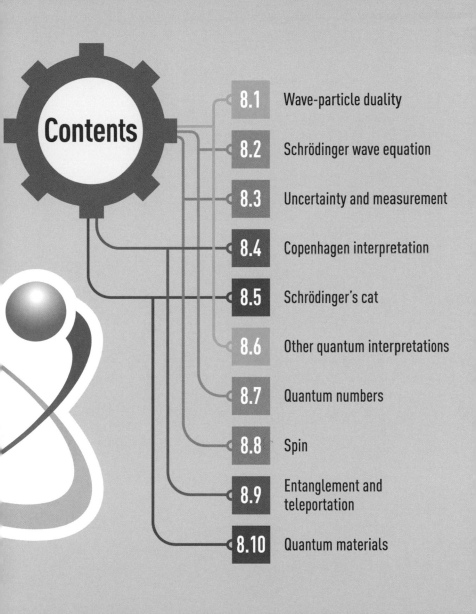

8.1 Wave-particle duality

Light behaves like a wave and a particle, while subatomic particles have hidden wave-like properties.

At the scale of atoms and particles, the physics describing what's going on has some very strange implications. Among these is the idea of wave-particle duality.

In 1900, German physicist Max Planck introduced the idea that energy might be quantized (broken down into tiny packets). Planck intended the idea as a neat mathematical trick, but in 1905 Albert Einstein showed that energy quanta were, in fact, real. They correspond to the packets of light known today as photons (see Topic 4.1). Einstein used them to solve a long-standing puzzle about the photoelectric effect (see Topic 5.5) – namely, why do some materials emit electrons in faint blue light, but not under intense red light? The explanation is that the energy of individual photons depends on their colour and wavelength: red photons just don't carry enough energy to dislodge electrons, no matter how many of them there are.

In 1924, the French scientist Louis Victor de Broglie suggested that duality might extend further. In a daring hypothesis, he showed that particles, such as electrons, must have an associated wavelength relating to their momentum.

Strangest of all is the quantum variant of Thomas Young's double-slit experiment (see page 189). Even if photons go through the slits one at a time, they still build up a pattern of lines characteristic of the interference of waves.

Electron microscopes can see inside human cells because electrons have a smaller wavelength than visible light.

The electron microscope relies on wave-particle duality: the fact that electrons have an associated wavelength much smaller than those of light photons, allows extreme magnifications to be reached, as in this close-up view of an ant.

8.2 Schrödinger wave equation

The properties of a particle cannot be calculated precisely, but are best described in probabilities by the mathematics of waves.

While Newton's laws of motion lie at the heart of classical physics, the Schrödinger wave equation is the linchpin of the quantum world.

Louis Victor de Broglie had suggested that every particle has an associated wavelength, and in 1926 Erwin Schrödinger took this idea further. If electrons could behave like waves, then surely he could describe their behaviour using a wave equation just like the one James Clerk Maxwell had used for electromagnetism (see page 87).

The solution of the wave equation is called the **wave function**. In classical wave theory, the wave function would tell you the shape of the wave at a particular time. But it cannot reveal a particle's location at a given time. Instead, it gives you the probability that the particle will be found in a particular spot.

Schrödinger developed his equation while spending winter in an isolated mountain cabin.

Like the Heisenberg uncertainty principle (see Topic 8.3), the wave function deals with probabilities rather than certainties. Despite this, Schrödinger believed that it did describe some kind of physical wave (though Niels Bohr and others disagreed – see Topic 8.4). Regardless of its true interpretation, the wave equation is enormously useful: it explains various physical phenomena including radioactive decay (see opposite) and the emission spectrum of hydrogen (see Topic 5.6).

Explaining radioactive alpha decay

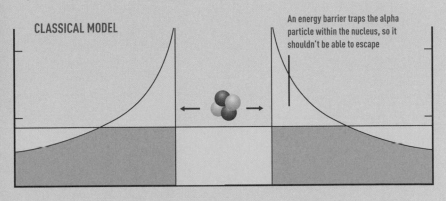

CLASSICAL MODEL

An energy barrier traps the alpha particle within the nucleus, so it shouldn't be able to escape

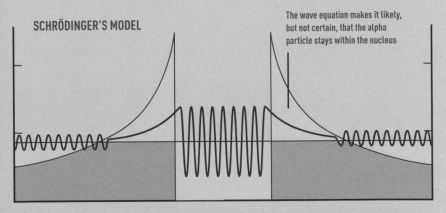

SCHRÖDINGER'S MODEL

The wave equation makes it likely, but not certain, that the alpha particle stays within the nucleus

In classical physics it should be impossible for an alpha particle to escape an unstable atomic nucleus and undergo alpha decay. But if the particle is described by a wave equation it has a small chance of tunnelling through the energy barrier and appearing on the other side.

8.3 Uncertainty and measurement

Some properties of quantum systems are inherently linked: the more accurately we pin down one, the less we know about the other.

One of the best known quantum ideas is the uncertainty principle developed by Werner Heisenberg. It describes what happens when you measure a particle's properties.

Heisenberg had been developing his own version of quantum mechanics, which produced identical results to the Schrödinger wave equation (see Topic 8.2). He realized that pairs of particular properties like a particle's position and momentum could not be measured accurately.

In 1927, Heisenberg published his idea, which showed that it was not possible to know both a particle's position and its momentum at the same time. If you are sure precisely where a particle is, you will not be able to determine its momentum with any degree of certainty.

The idea relates to Schrödinger's wave equation. The places at which the wave function is most intense tells you where, on average, most particles will be. The uncertainty is the range of possible positions around that location.

The uncertainty principle affects time and energy, allowing nature to conjure up short-lived virtual particles out of nothing.

The principle explains why atoms don't collapse. The opposite charges of protons and orbiting electrons should attract each other, but the uncertainty in position prevents electrons spiralling too close to the nucleus because their speeds would get too high. Nature is 'fuzzy' at this scale.

Uncertainty in action

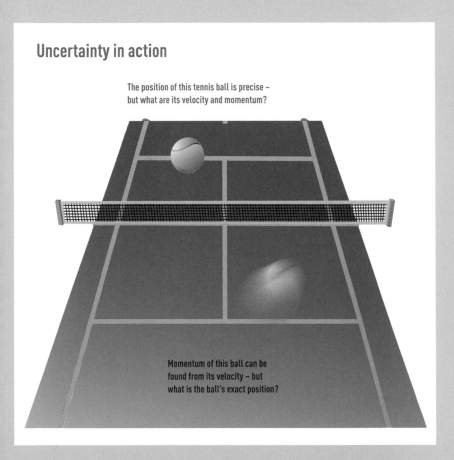

The position of this tennis ball is precise –
but what are its velocity and momentum?

Momentum of this ball can be
found from its velocity – but
what is the ball's exact position?

One way of thinking about the uncertainty principle is to consider a moving object such
as a tennis ball. Depending on the measurement taken, you can determine its precise
position, or its exact momentum, but you can never find both simultaneously.

8.4 Copenhagen interpretation

The uncertainties of the quantum world collapse into certainty on interaction with the larger universe.

With new ideas about the quantum world emerging, a group of physicists set out to explain how they related to the real world around us. The Copenhagen interpretation was born.

The name refers to the Danish capital city, where pioneers Niels Bohr, Werner Heisenberg and colleagues worked in the 1920s. The picture presented by their equations had probabilities, as shown by Schrödinger's wave equation; it was one of uncertainty, as shown by Heisenberg; and it had an idea Bohr called complementarity.

Complementarity related the quantum world to the everyday world by saying that waves and particles are complementary. An object could not behave as both a wave and a particle at the same time. Laboratory experiments demonstrated either their wave or particle nature, depending on the situation (see opposite).

What quantum theory showed, they said, was that the properties of physical systems aren't definite before they are measured. At that point, the wave function collapses to reveal the world that is being observed.

Einstein voiced his objections to the Copenhagen interpretation with his famous adage 'God does not play dice'.

The Copenhagen interpretation remains the dominant way of thinking about quantum mechanics, though today there are several other interpretations (see Topic 8.6).

Wave or particle?

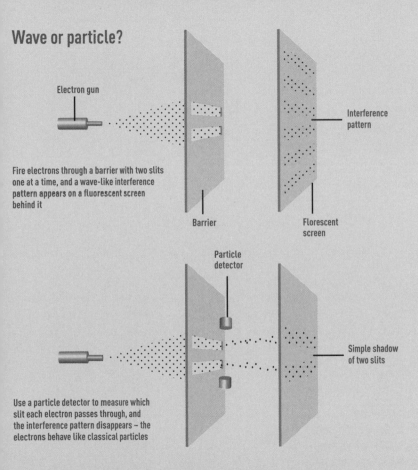

Electron gun

Fire electrons through a barrier with two slits one at a time, and a wave-like interference pattern appears on a fluorescent screen behind it

Barrier

Florescent screen

Interference pattern

Particle detector

Use a particle detector to measure which slit each electron passes through, and the interference pattern disappears – the electrons behave like classical particles

Simple shadow of two slits

The quantum double-slit experiment is analogous to the one used by Thomas Young to prove the wave nature of light (see Topic 2.4), but its strange result can only be explained by an idea such as the Copenhagen interpretation.

8.5 Schrödinger's cat

Adherence to the Copenhagen interpretation gives rise to an absurd situation.

The Copenhagen interpretation (see Topic 8.4) introduced the idea that particles existed in a superposition of possible states, only collapsing into a single state when they're measured. But one pioneer of quantum theory didn't believe this would apply in the everyday world, and came up with a famous thought experiment to highlight the problem.

Erwin Schrödinger believed his wave equation really did describe an intrinsic property of matter itself. He imagined a situation in which a quantum event, such as the decay of a radioisotope, was linked to a large object – a cat. The decay would trigger the release of poisonous cyanide, killing the cat.

The whole (imaginary) experiment would be concealed inside a box. And since the decay would be random, there would be no way of knowing if the cat were alive or dead at any given moment. In the Copenhagen interpretation, the cat would be both alive and dead at the same time until the box was opened, collapsing the wave function. Only then could we determine whether or not the cat was still breathing.

Schrödinger developed his cat model in response to discussions with Einstein about the absurdity of superposed states.

Schrödinger thought this contradictory state of affairs was ridiculous, and that it couldn't describe how nature really worked. The experiment sparked decades of debate, with new ideas emerging to challenge the Copenhagen interpretation (see Topic 8.6).

The cat in the box

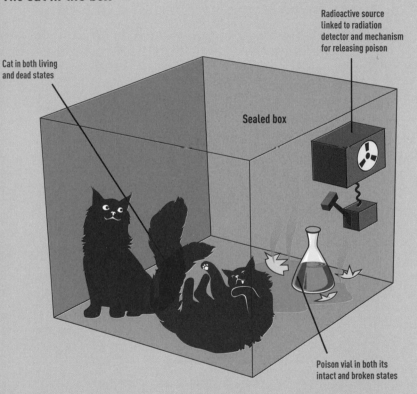

Radioactive source
linked to radiation
detector and mechanism
for releasing poison

Cat in both living
and dead states

Sealed box

Poison vial in both its
intact and broken states

According to the strictest version of the Copenhagen interpretation, the
entire system – the radioactive source, the poison vial and the cat – remains
in a superposition of two possible states until an observer opens the box.

8.6 Other quantum interpretations

How exactly does the quantum wave function affect the world we see? Physicists have come up with many different solutions to this philosophical problem.

Alongside the Copenhagen interpretation sit many other ways of explaining why the wave function collapses when a measurement is made. In doing so, they provide different philosophical ideas about reality itself.

▪ In the **consistent histories** approach, the wave function is merely maths. Quantum mathematics can be used to work out what will happen to large-scale objects as well as particles. Its job is not to describe the physical state of the particle, but to map the probability of various outcomes or histories.

▪ The **ensemble interpretation** says the wave function does not describe a single particle, but a huge array or ensemble of identical systems. The wave function is merely responsible for guiding us to our particular system.

▪ The **many worlds interpretation**, developed by Hugh Everett III in the 1950s, says that our universe is constantly branching into alternate universes – one for each possible outcome of an event. Schrödinger's cat would actually be both alive and dead at the same time, just in different universes.

Some physicists eschew the whole idea of trying to interpret how quantum theory describes reality. They treat the wave function as pure math, an approach described as 'shut up and calculate!'

The ensemble interpretation was Einstein's favourite approach to quantum physics.

Schrödinger's (many) cats?

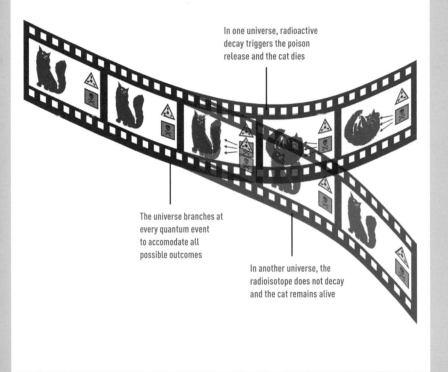

In one universe, radioactive decay triggers the poison release and the cat dies

The universe branches at every quantum event to accomodate all possible outcomes

In another universe, the radioisotope does not decay and the cat remains alive

In order for the many worlds interpretation of quantum physics to be true, we would have to live in a multiverse of infinite dimensions, allowing each branching universe to remain forever separated from all the others.

8.7 Quantum numbers

Various subatomic properties fall into quantized units that help us understand the structure of atoms.

One important and strange aspect of quantum physics is that many subatomic properties are quantized – that is, they can only be whole or half-numbers. This is strange, because in the world of classical physics, properties like momentum can have any value.

Four of the most important quantum numbers describe an electron's properties within an atom. They are known as the principal, azimuthal, magnetic and spin quantum numbers (denoted by n, l, m and s respectively). The larger the principal number, the further an electron is from the nucleus of an atom.

In 1925, Wolfgang Pauli discovered that, for a large group of particles known as **fermions** (which includes electrons), no two particles in a system, such as an atom, could share the exact same set of quantum numbers. This exclusion principle explains the shell-like arrangement of electron orbits (see Topic 5.4). The principal quantum number describes the main shell within which an electron lies, while the azimuthal and magnetic numbers define the subshells. Finally, the spin quantum number explains why two electrons can occupy each subshell (see Topic 8.8).

Patterns in quantum numbers can be used to describe a variety of chemical properties of elements.

Fermions, bosons and energy levels

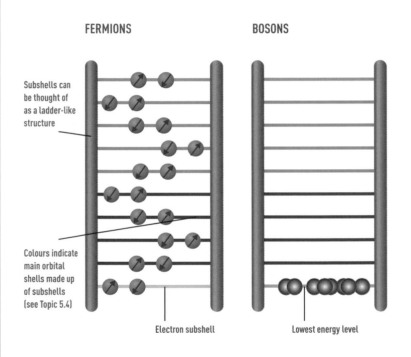

FERMIONS

BOSONS

Subshells can be thought of as a ladder-like structure

Colours indicate main orbital shells made up of subshells (see Topic 5.4)

Electron subshell

Lowest energy level

The Pauli exclusion principle ensures that each energy level in an atom is occupied by a maximum of two electrons with opposite spins (see Topic 8.8), and thus creates an atom's overall electron structure. Bosons suffer no such restrictions and tend to fall into the lowest possible energy state.

8.8 Spin

There is a fundamental dividing line between the two different types of particles in the universe.

Spin is one of the most important but confusing aspects of the quantum realm. It subdivides all particles in the universe into two fundamental types with very different properties.

Despite its name, quantum spin is not simply a particle's rotation around its axis. Instead, it seems that subatomic particles don't rotate at all, but their electric charge rotates within them, generating a tiny, but measurable, magnetic field as it does so.

Spin is quantized, occurring in discrete values – that is, 0, whole or half integers. It's also conserved: particle interactions such as radioactive decay always retain the same total amount of spin before and after. And, on the level of individual particles, spin can be reversed, but never increased or decreased. Particles with half-integer spins are known as fermions, and comprise all the familiar matter particles in the universe. The electron, for example, has a spin of ½. Crucially, fermions obey the Pauli exclusion principle (see Topic 8.7), which means they cannot be in the same place with the same spin.

With spins of 0 or a whole number, bosons observe no such restrictions. The force-carrying particles of the universe (see Topic 9.2), they obey a set of rules called Bose-Einstein statistics, which have strange consequences of their own (see Topic 8.10).

Spintronics is a field of electrical engineering that makes use of spin to improve devices such as magnetic hard disk drives.

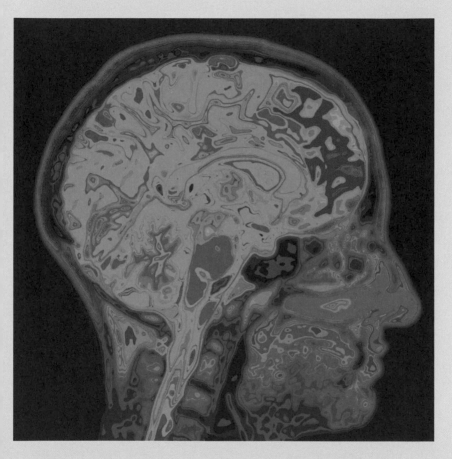

Magnetic resonance imaging (MRI) is a medical technique that temporarily disrupts
the spin of hydrogen nuclei (protons) in the body's internal water, and then studies the
radio signals emitted as they relax back into their natural state.

8.9 Entanglement and teleportation

Quantum theory suggests it's possible to create a truly secure means of sending messages.

Einstein called entanglement 'spooky action at a distance'. It was the idea that two particles could be connected in such a way that changing one would also change the other, no matter how far apart they were.

Entanglement arises from the Pauli exclusion principle (see Topic 8.7). Because no two fermion particles in a system can share the same quantum numbers, it's possible to create **entangled** pairs in which measuring the properties of one fermion (the spin of an electron, say) inevitably reveals the properties of the other. Now imagine separating an entangled pair while allowing their states to remain unmeasured. If used to secure information, it would act like a tamper-proof lock. Measuring one particle would automatically change the other because they have always been entangled, so trying to hack into the system would simply destroy the lock.

Quantum encryption holds the promise of a truly secure Internet – entangled pairs have so far been separated by up to 102 km (63 miles). But what if data were encoded in the spin states themselves? It could be teleported across the universe, faster than the speed of light. Entanglement seems to defy logic, but because it is not possible to select the particle's state, this prevents it from being used to communicate information faster than light speed. Yet it still has some amazing potential applications.

The term 'entanglement' was first used in this context by Erwin Schrödinger.

Quantum teleportation

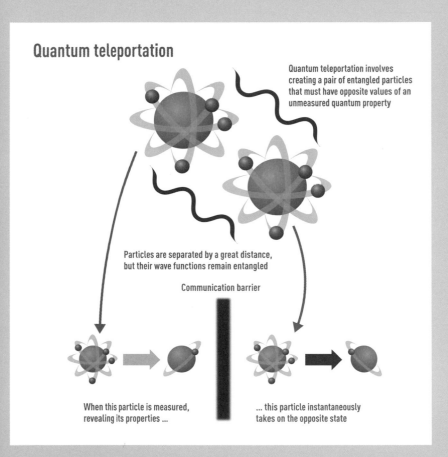

Quantum teleportation involves creating a pair of entangled particles that must have opposite values of an unmeasured quantum property

Particles are separated by a great distance, but their wave functions remain entangled

Communication barrier

When this particle is measured, revealing its properties ...

... this particle instantaneously takes on the opposite state

When entangled particles are separated by a large distance or some other communication barrier, measuring the property in one of the pair will instantaneously resolve it in the other – it's not necessary for a signal to travel between the two.

8.10 Quantum materials

The laws of quantum physics can create strange materials – from superdense matter to frictionless fluids.

The physics of the quantum world seem strange and esoteric, but they explain the most extreme forms of material in space as well as those created on Earth.

One of the densest forms of matter occurs in a neutron star – a star that's exploded and collapsed into a ball the size of a city. It's only prevented from collapsing further by so-called degeneracy pressure between particles (fermions), which are prevented from falling into a single low-energy state by the Pauli exclusion principle (see Topic 8.7).

Another type of quantum material is the superfluid, or Bose-Einstein condensate. This is made from bosons (particles with integer spins). Under extremely low temperatures close to absolute zero, all their atoms fall into the lowest possible quantum state. They combine to form a super atom with an overall whole-number spin. A superfluid can be made from the gas helium, and can flow as a liquid without encountering friction.

Something similar can happen to electrons (fermions) inside certain materials. At low temperatures they form boson-like **Cooper pairs** and overcome the exclusion principle, falling into identical energy states and then flowing through a conductor without resistance – an effect called **superconductivity**. Superconducting magnets are hugely powerful and have been used in particle accelerators.

A neutron star just 20 km (12 miles) in diameter would have 1.4 times the mass of the Sun.

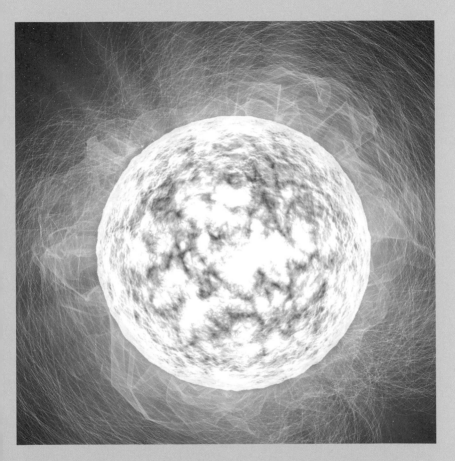

An artist's impression of a neutron star. Such stars are normally too faint to detect, but young neutron stars, recently formed in supernova explosions, frequently concentrate all their radiation in two narrow magnetic beams, creating a cosmic lighthouse called a pulsar.

PARTICLE PHYSICS

H ow deep does the structure of matter go? For the physicists of the late-19th century, atoms appeared to be nature's ultimate building blocks. But then the discovery of the electron, proton and neutron revealed the atom's inner structure. And that, however, was just the beginning.

In their quest to understand how the properties of subatomic particles arise, physicists soon revealed another layer of underlying structure. It became clear that protons and neutrons could be subdivided again, into particles called quarks. Only the electron remains (we think) indivisible.

Particle physics is the cutting edge of modern science, concerned with identifying the truly elementary building blocks of the universe, and the fundamental forces that allow them to interact. The science wrestles with

Continues overleaf

the consequences of quantum theory, and the problem that such particles are extremely hard to extract from atoms. Not only that, but they disintegrate almost instantaneously when removed, making them difficult to study. To overcome this, scientists use particle accelerators, the largest machines ever built, to create high-energy environments in which elementary particles can briefly be observed outside of atoms.

Our tour of this strange world begins with a review of the elementary particles in the current standard model of particle physics, the four forces that govern them and the techniques used to study them. We then look at three of the forces in detail – the fourth, gravity, is covered in Chapter 10. Finally, we look at theories and new discoveries that may some day provide us with a fuller picture of the ways in which the universe works.

Contents

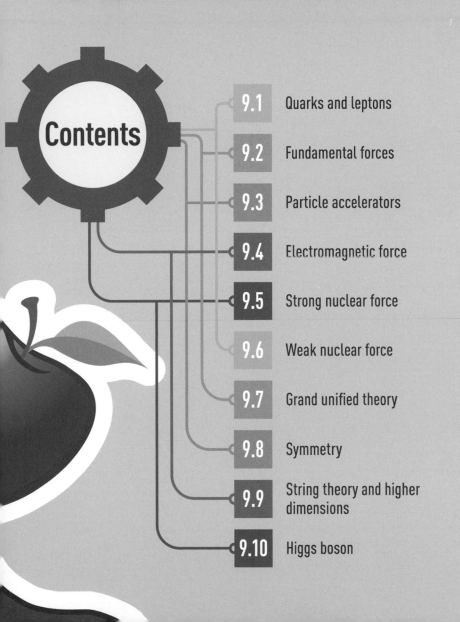

9.1 Quarks and leptons

Two major classes of particles make up all the matter in the universe – quarks, found for example in the atomic nucleus, and light leptons such as the electron.

The 20th century saw the discovery of new subatomic particles beyond the familiar proton, electron and neutron. The discoveries led to a new way of explaining matter.

In the wake of the advances in nuclear physics made during World War II, and early studies of cosmic rays (high-energy particles from deep space), physicists in the early 1960s found themselves faced with a chaotic 'particle zoo'. Measuring the motion of particles through magnetic fields revealed information about their mass and electric charge, and uncovered a great many new particles. In general, these now seemed to divide into a multiplicity of heavier particles (hadrons) and just a few lighter particles (leptons).

To explain the huge variety of hadrons, Murray Gell-Mann and George Zweig proposed a deeper level of particle altogether – the quark. The initial idea was that three **flavours** of quark could combine in pairs or triplets to make hadrons. For example, a proton (with an electric charge of +1) is made up of two **up quarks** (each with charge of $+\frac{2}{3}$) and one **down quark** (with charge of $-\frac{1}{3}$). The existence of quarks has been confirmed in experiments, and there are now known to be six flavours in all.

Today, quarks form one major branch of the standard model of particle physics. As fermions (see Topic 8.7), they sit alongside leptons as the matter particles of the universe.

The word quark features in James Joyce's novel *Finnegans Wake*.

Particles of the standard model

Increasing mass --------------->

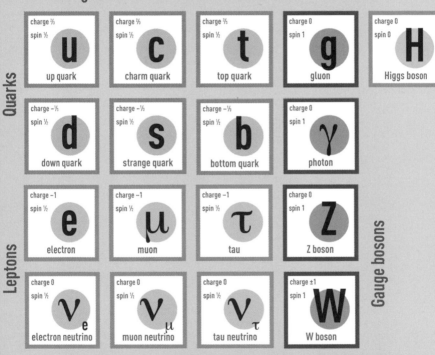

The standard model of particle physics suggests there are just 12 matter particles (though each has an antimatter equivalent), four force-carrying gauge bosons and the Higgs boson.

9.2 Fundamental forces

Just four forces are responsible for all the interactions between matter so far discovered.

For centuries, it seemed that gravity and electromagnetism could explain everything we experience in the world. But as physicists learned more about the subatomic realm, it became clear that additional forces were at work.

According to the standard model developed in the 1960s, there are four **fundamental forces**: electromagnetism, the strong nuclear force, the weak nuclear force and gravitation.

- Electromagnetism affects all particles with electric charge. It encompasses both electricity and magnetism.

- The strong nuclear force affects only hadrons – quarks and the wide variety of particles composed of quarks.

- The weak nuclear force affects both quarks and the six leptons, of which the electron is one.

- Gravitation is the outsider of the four: it is extremely weak and only becomes significant when large quantities of matter are involved. Gravitation, however, is best understood in terms of general relativity (see Chapter 10).

At least three of the forces (electromagnetism and the two nuclear forces) are transmitted or mediated by exchanging another type of particle called a **gauge boson** (see Topic 9.1).

Some scientists have argued that we may need a fifth force in order to explain all the observed features of our universe.

The four forces

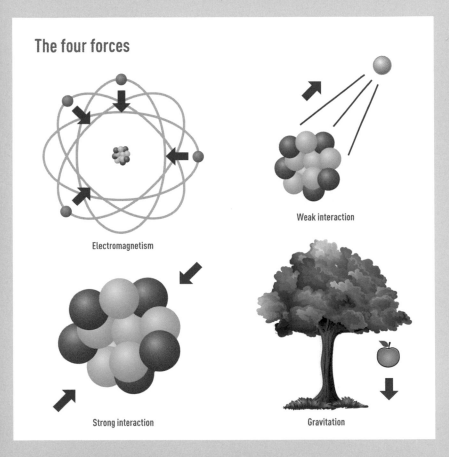

Electromagnetism

Weak interaction

Strong interaction

Gravitation

The four fundamental forces of nature hold all matter together with different degrees of strength and over different distances, but can they be reconciled with one another, as physicists hope (see Topic 9.7)?

9.3 Particle accelerators

Physicists build huge machines capable of creating high-energy environments so they can study subatomic particles.

Modern research into particle physics frequently involves using particle accelerators – enormous machines used to create particles that don't occur naturally on Earth. They are like time machines, recreating conditions in the early universe.

Particle accelerators use linear or ring-shaped tracks lined with electromagnets to accelerate charged particles to huge speeds, forcing them to collide. The energy that comes out of a collision is the sum of the energy of the colliding particles.

Energy is the key to particle accelerators because, in Einstein's equation $E = mc^2$ – energy is equivalent to mass. So the bigger the energy, the more massive the particles produced in the collision. Using particle accelerators, scientists are able to see particles that wouldn't normally exist on Earth – particles like those that started in the big bang of enormous energy.

One of the applications of particle accelerators in physics is to investigate whether the particles and forces invented as part of theories really exist. For example, finding evidence for quarks (see Topic 9.1) in the 1970s confirmed that the theory was broadly correct. The greatest achievement of particle physics so far is probably the discovery of the elusive Higgs boson in 2012 (see Topic 9.10). It was found using the Large Hadron Collider (LHC), located on the border between France and Switzerland.

The Large Hadron Collider's main tunnel is 27 km (17 miles) long, and kept in vacuum conditions comparable to outer space.

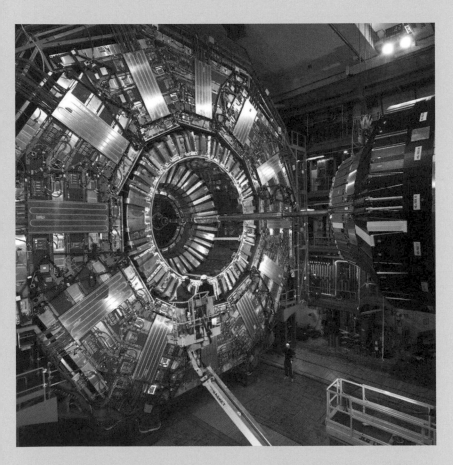

The huge Compact Muon Solenoid (CMS) experiment at the Large Hadron Collider
(LHC) is a particle detector chamber some 21.6 m (71 ft) long. It played a key role
in the discovery of evidence for the Higgs boson.

9.4 Electromagnetic force

This force between charged particles is carried by photons of light – messengers that are spontaneously created and destroyed.

The electromagnetic force in physics doesn't relate to electric and magnetic fields alone. The theory encompassing it also describes a huge range of phenomena and was called the jewel of physics by its co-creator Richard Feynman.

In 1873, the Scottish physicist and mathematician James Clerk Maxwell proposed the idea that light was an electromagnetic wave (see Topic 4.1). All the phenomena of electricity, magnetism and light could be described using this single theory. But when quantum theory was developed for atoms and subatomic particles, Maxwell's concept needed modification to show how these particles interacted with light. Quantum electrodynamics (QED) had its beginnings in the 1920s, but was fully developed in the 1940s by a trio of physicists that included Feynman.

In QED, and other related theories, force is transmitted by an exchange of particles called gauge bosons (see Topic 9.1). These are massless photons of electromagnetic radiation similar to light (Topic 4.1). However, such photons are not necessarily detectable as they are created and destroyed in the process. One of the strange predictions of QED is the role of 'virtual' particles that can wink in and out of existence in fractions of a second thanks to Heisenberg's uncertainty principle (see Topic 8.3). The idea of virtual particles might seem surprising, but it is supported by the curious phenomenon known as the Casimir effect (opposite).

Physicists model electromagnetic force interactions using a simple but powerful tool called a Feynman diagram.

Virtual particles and the Casimir effect

The existence of virtual particles is confirmed by the
Casimir effect – an otherwise inexplicable force between
parallel plates in a vacuum

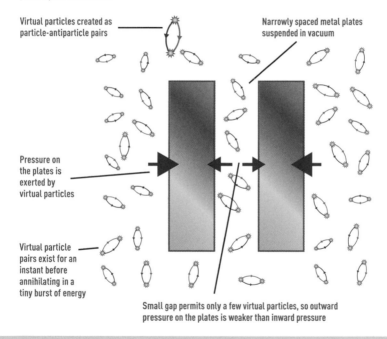

Virtual particles created as
particle-antiparticle pairs

Narrowly spaced metal plates
suspended in vacuum

Pressure on
the plates is
exerted by
virtual particles

Virtual particle
pairs exist for an
instant before
annihilating in a
tiny burst of energy

Small gap permits only a few virtual particles, so outward
pressure on the plates is weaker than inward pressure

Heisenberg's uncertainty principle means that it's possible to create
pairs of virtual particles capable of carrying forces by borrowing
tiny amounts of energy out of vacuum for very brief periods of time.

9.5 Strong nuclear force

The force that binds quarks together to create protons and neutrons also binds the atomic nucleus.

When physicists discovered more about what was inside the atom, it threw up a problem: why don't atoms spontaneously explode? The answer was a hitherto undiscovered force holding them together.

The atomic nucleus contains only neutrons, which have no electric charge, and positively charged protons. Positive charges repel each other, so it follows that every atom should blow apart. The fact that they don't is down to the strong nuclear force, acting like glue to bind them together.

The strong force operates between protons and neutrons, and also inside them, holding together the three quarks within each one. It is the most powerful of all the fundamental forces but it also operates over the shortest range, equivalent to roughly the diameter of a proton. Like other forces, it operates by exchanging so-called messenger particles. The particles carrying the strong force are **gluons**. Evidence for their existence came in 1979.

Strong force is carried between protons and neutrons by a messenger particle called the pion, itself consisting of two virtual quarks.

Following the success of the QED model in explaining electromagnetism (see Topic 9.4), physicists developed **quantum chromodynamics** or QCD to describe the strong force. This assigns quarks with a property known as colour, which governs how they bind together. Only particular combinations of colour are possible, explaining why certain quarks are observed as hadron particles in particle accelerators.

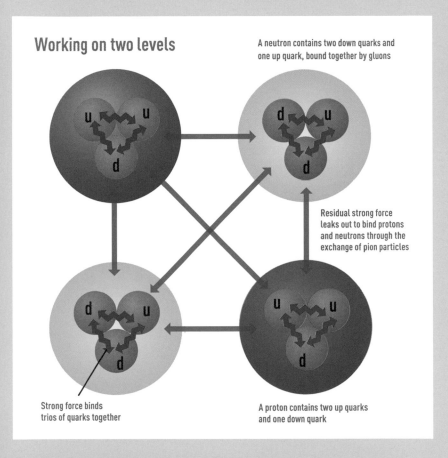

Working on two levels

A neutron contains two down quarks and one up quark, bound together by gluons

Residual strong force leaks out to bind protons and neutrons through the exchange of pion particles

Strong force binds trios of quarks together

A proton contains two up quarks and one down quark

The nuclear force is strongest between the quarks inside each proton or neutron, where it is carried by gauge boson particles called gluons. In a much weaker residual form, it also binds protons and neutrons.

9.6 Weak nuclear force

A second force in the atomic nucleus plays a key role in radioactive decay.

The weak nuclear force lives up to its name: it's weaker than all the other forces except gravity. In fact, it plays a greater role in things falling apart or decaying than pulling things together.

The weak force was needed to explain beta decay (see Topic 7.1) – when a neutron decays into a proton and an electron, and the electron is emitted from the nucleus. Physicists knew that the strong force couldn't be responsible, as it did not act on electrons, while the electromagnetic force did not act on neutrons, and gravity was too weak. It was Italian physicist Enrico Fermi who published a theory outlining the weak force in 1934, suggesting that it converted protons into neutrons and neutrons into protons, at the same time emitting electrons or positrons, as well as neutrinos, from the nucleus.

Later, physicists would search for the particle that carried the weak force and find, not one, but three: the W^+, W^- and Z bosons. The two W particles are at work when the particles involved in interactions swap charges, while the Z is neutral. Their high mass explains why the force only works over a very short range. Intriguingly, the weak force is the only one that violates some of the symmetries of nature (see Topic 9.8).

The existence of the W and Z bosons was only confirmed in the 1980s.

We have reasons to be thankful for the weak force. It gives rise to the positrons used to diagnose diseases in PET scanners and it governs fusion reactions in the Sun. If the weak force were not so weak, the Sun would have stopped burning long ago.

Nuclear transformations

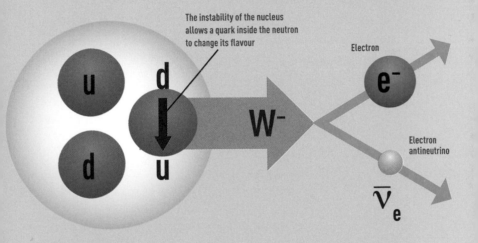

The instability of the nucleus
allows a quark inside the neutron
to change its flavour

Radioactive beta decay (in which a neutron transforms into a proton) actually involves
a single down quark changing into an up quark and emitting a W⁻ gauge boson in the
process. The W⁻ particle then rapidly decays into an electron and its antineutrino.

9.7 Grand unified theory

Theoretical physicists dream of a single superforce in which all fundamental forces were once combined.

One aim of modern theoretical physics has been to identify similarities in the behaviours of the four fundamental forces. The ultimate ambition is to formulate a single theory that describes the entire universe, a **theory of everything**.

The task of uniting gravity with the other three forces in a single theory of everything (see Topic 9.9) is extremely difficult. However, the electromagnetic, strong and weak forces show tantalizing similarities in their behaviours and this has led physicists to work on a **grand unified theory** (GUT) to unite them, instead.

Evidence from particle accelerators has already shown that, above certain energy levels, the electromagnetic and weak forces can be combined in a single **electroweak** interaction, and theoreticians hope that the strong force can ultimately be added to produce an **electronuclear** force.

If a GUT could be found, its discovery might also bring a solution to one of the biggest questions about the early universe. Within a fraction of a second of creation, when energies were still unimaginably high, the infant cosmos went through an inexplicable, dramatic burst of expansion known as **inflation**. Some cosmologists speculate that inflation might have been driven by energy released when the forces we experience today **froze out**, or separated from, the primeval superforce the initiated the big bang.

The acronym GUT was first coined in 1978 by researchers at the CERN particle physics lab.

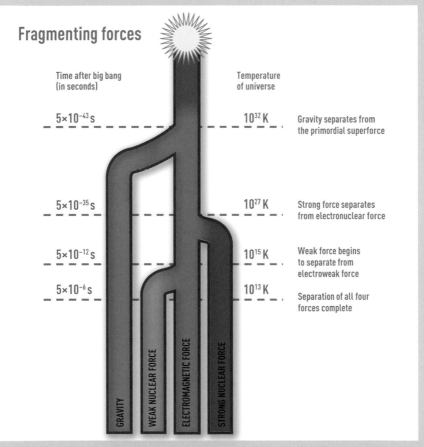

Fragmenting forces

Time after big bang
(in seconds)

Temperature
of universe

5×10^{-43} s 10^{32} K   Gravity separates from
the primordial superforce

5×10^{-35} s 10^{27} K   Strong force separates
from electronuclear force

5×10^{-12} s 10^{15} K   Weak force begins
to separate from
electroweak force

5×10^{-6} s 10^{13} K   Separation of all four
forces complete

GRAVITY   WEAK NUCLEAR FORCE   ELECTROMAGNETIC FORCE   STRONG NUCLEAR FORCE

According to the best theories, today's four forces separated out of a primordial superforce
within the first millionth of a second after the big bang. Particle accelerators aim to recreate
conditions from this time in the hope of understanding how the forces were once unified.

9.8 Symmetry

How do the fundamental forces behave when the properties of particles are changed?

When a mathematical shape is symmetrical, it looks like a mirror image either side of a line of symmetry. In physics, symmetry applies to other properties and it has far-reaching implications for our understanding of nature.

The reflection of a symmetrical shape in a mirror is called a **transformation**. Particles and forces can also undergo transformations to their properties. A force is said to have symmetry if it is the same before and after a transformation.

There are three types of symmetry affecting fundamental forces and elementary particles: **charge** (C) sees electrical charges reversed; **parity** (P) flips the orientation and spin of a particle; and **time** (T) reverses the direction of time. A particle is symmetric if you perform the same transformation twice and get back to what you started with.

Some processes have a single type of symmetry; others are symmetrical under a combination of transformations. For example, CP symmetry involves reversing both charge and parity. When physicists investigated symmetry, they discovered that the weak nuclear force (see Topic 9.6) violates not just during processes such as beta decay, but CP as well, which was thought to be impossible. CP violation could explain why we are here, as it might have been responsible for the creation of more matter than antimatter (see Topic 7.9) in the big bang.

Supersymmetry is a hypothetical symmetry between matter particles and force carriers, but it remains unproven.

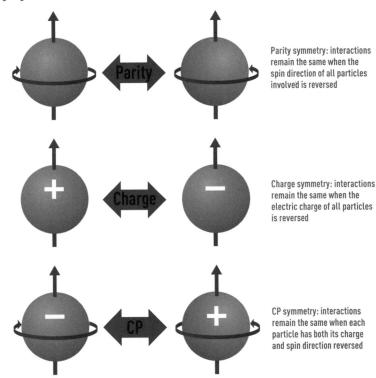

Key symmetries in fundamental forces

Parity symmetry: interactions remain the same when the spin direction of all particles involved is reversed

Charge symmetry: interactions remain the same when the electric charge of all particles is reversed

CP symmetry: interactions remain the same when each particle has both its charge and spin direction reversed

While gravity, electromagnetism and strong nuclear force interactions are all charge-, parity- and CP-symmetric, the weak nuclear force breaks all these symmetries as well as time-symmetry. Like all forces, however, it obeys the CPT symmetry in which all three properties are reversed.

9.9 String theory and higher dimensions

Popular theories suggest that particle properties might arise from string-like vibrations.

The best-known candidate for a potential theory of everything (see Topic 9.7) is **string theory**. It's a deceptively simple concept, yet it has huge implications, including the existence of extra dimensions we can't see.

The central premise of string theory is that subatomic particles are in fact vibrating loops of energy. These strings have a tendency to fall into harmonic oscillations like those of a musical instrument (see Topic 2.2). The precise harmonics of a string vibrating determine the value of its quantum numbers (see Topic 8.7), neatly explaining why these properties jump between discrete values instead of varying continuously.

Despite the simplicity of the basic idea, finding a working version of string theory that conforms to the real universe has proved difficult. So-called **superstring** theories solve many problems, but require at least ten dimensions. Since our everyday experience involves just four dimensions – three of space and one of time – where are the others? One possibility is that they are all around us, but curled up so tightly that we simply cannot detect them. Alternatively, if one (or more) of the extra dimensions is not curled up, it could create a five-dimensional hyperspace through which our universe and others drift as membrane-like structures called branes (see opposite).

Strings are far too small for us to observe directly in the near future, but particle accelerators might provide evidence for higher dimensions.

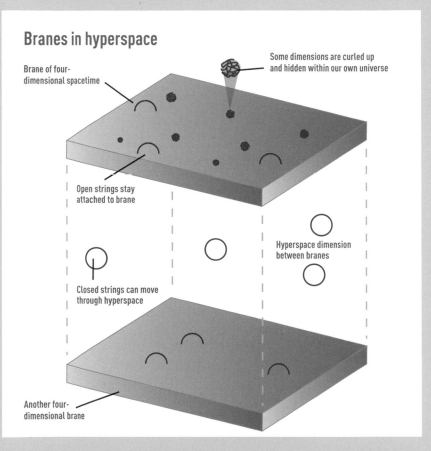

Branes in hyperspace

Brane of four-dimensional spacetime

Some dimensions are curled up and hidden within our own universe

Open strings stay attached to brane

Closed strings can move through hyperspace

Hyperspace dimension between branes

Another four-dimensional brane

For simplicity, this diagram visualizes four-dimensional branes as planes separated by a third dimension of hyperspace. Some think that a collision between such branes could have triggered the big bang that created our universe.

9.10 Higgs boson

No particle is created with mass – instead, mass stems from the activity of an illusive extra field, the Higgs field.

The **Higgs boson** was proposed in 1964 by Peter Higgs, as a side effect of one of the biggest puzzles in physics – what is mass and where does it come from?

The existence of mass as a property became a focus for research following the discovery of the W and Z bosons that carry the weak nuclear force in the 1980s (see Topic 9.6). Unlike the force-carrying photon and gluon, these particles were not massless. On the contrary, they were extremely heavy. They have 100 times more mass than the proton – more even than the nucleus of a copper atom.

The standard model could not explain the discrepancy, but a solution did come to light. Back in the 1960s, British physicist Higgs had developed a mathematical framework that generated masses for the particles. It involved a field that permeated the universe. When particles interacted with this field they gained mass. Different particles would undergo different interactions, explaining the variety of masses and why certain particles remained massless.

In popular media the Higgs boson is often known as the God particle.

Although the Higgs field can't be detected directly, a disturbance of the field would form a quantum particle: a boson with zero spin and zero charge. This boson also had a high mass, making it beyond the reach of particle accelerators until the Large Hadron Collider (LHC) came along. It's likely that the Higgs was finally spotted by the LHC in 2012.

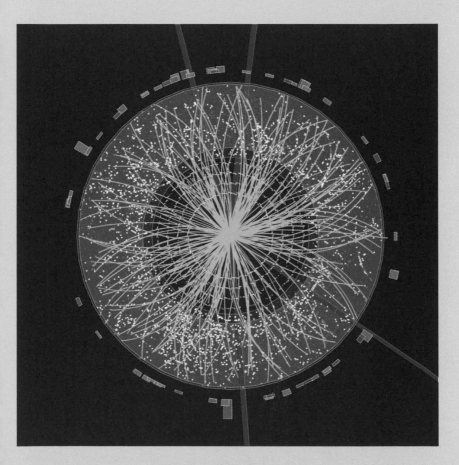

A plot of data from a 2012 run of the Large Hadron Collider's ATLAS
experiment shows the decay of a Higgs-like particle into four muons
(red tracks) amid a complex burst of quark-based hadron particles.

RELATIVITY

n 1905, a 26-year-old German physicist by the name of Albert Einstein, then working in the Swiss Patent Office, published no fewer than four revolutionary scientific papers. One provided clinching statistical evidence for the existence of atoms, while another kick-started the quantum revolution (see Chapter 8).

The two remaining papers are the ones for which Einstein is best remembered, however. Together, they outlined a new, revolutionary approach to physics, rewriting the laws of mechanics from the ground up in order to take account of the fixed speed of light.

Einstein's theory of special relativity did not spring out of the blue – already in the late-19th century several other physicists were working to resolve significant problems with the classical view of the universe. None, however, dared to do what Einstein did, in

Continues overleaf

discarding the previous rules completely and rebuilding everything from first principles. A decade later, he followed up with the even more ambitious general theory, which explained how the presence of large masses distorts time and space around it to create gravity. The predictions of relativity often run counter to common sense, but in this case it's common sense that's wrong: relativity has been proved correct in experiment after experiment.

Our overview of this intriguing field begins with a brief look at the problems Einstein was seeking to correct, before turning to special relativity and some of its implications. In the latter half of the chapter we look at the general theory, its predictions and some of the extraordinary objects and behaviours that can arise in Einstein's fascinating universe.

Contents

10.1 Newton's failures

The problems with Newtonian physics included the fixed speed of light and the action of gravity.

By the late-19th century, cracks had appeared in the classical physics ushered in by Isaac Newton. A couple of problems that seemed minor opened the door to a major change in physics.

One problem was light. Maxwell's electromagnetic wave equation (see Topic 4.1) gave a value for the speed of light in empty space of around 300,000 km/s (186,000 miles per second). But what was light travelling through? The medium through which sound waves travel is air, so what was light's medium?

Prevailing opinion assumed the presence of an all-pervading 'luminiferous ether', but if this was true then light should behave like any other wave, appearing to move faster or slower depending on the relative motion of source and observer. Light, however, seemed to remain stubbornly at the same speed all the time. In 1887, scientists Albert Michelson and Edward Morley devised an ingenious experiment using interferometry (Topic 2.7) that should have detected even the tiniest change in the speed of light due to Earth's motion through the ether. It found nothing.

Another issue lay with gravity. Newton's law of universal gravitation was highly successful in explaining Kepler's laws of planetary motion (see Topic 1.5). However, Mercury's elliptical orbit around the Sun wobbled, or **precessed**, more quickly than could be explained by the influence of the other planets.

Ultimately both problems would only be solved by an entirely new approach to physics – relativity.

The strange behaviour of Mercury's orbit was once blamed on the influence of an undiscovered planet called Vulcan.

The Mercury problem

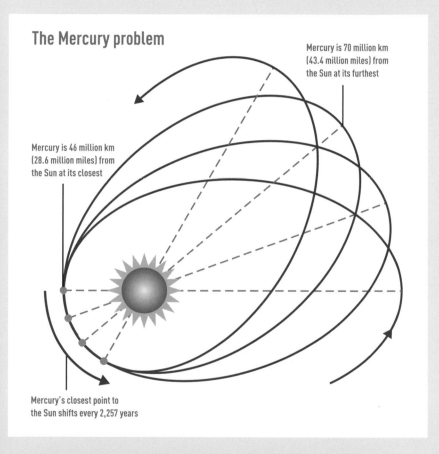

Mercury is 70 million km (43.4 million miles) from the Sun at its furthest

Mercury is 46 million km (28.6 million miles) from the Sun at its closest

Mercury's closest point to the Sun shifts every 2,257 years

Mercury's orbit is the most elongated of any in the solar system, and its orientation slowly wobbles or 'precesses' round the Sun. However, the rate of precession is 180 years faster than can be explained by the tug of the other planets' gravity alone.

10.2 Special relativity

Einstein showed that the constant speed of light does away with the idea of universal time.

James Clerk Maxwell's equations showed that light travelled at a constant speed, but it was Albert Einstein who showed that this had profound consequences.

Einstein thought that the laws of physics must be the same for all observers in motion, as long as the motion is steady or inertial (the principle of relativity). Crucially, all observers in **inertial reference frames** would measure the same speed for light.

This meant there was no universal time that everyone would agree upon. In other words, simultaneity was **relative** (see opposite). Imagine flashes of light at A and B and an observer centred between them. In the top example, the light flashes arrive simultaneously at the observer, who decides the flashes happened at the same time. In the bottom example, the entire platform is in motion. Unless the observer on board is aware of that, he will say that the flashes were not simultaneous. An observer in a different reference frame who perceives the motion, however, would argue that they were.

The principle of relativity was put forward by Galileo Galilei as early as 1632.

The beauty and confusion of special relativity lies in the fact that there is no privileged frame of reference from which to measure events – both observers are equally correct. The effects of this only become apparent when the speeds involved approach the speed of light, but relativity's effects stretch far beyond the world of abstract thought experiments.

Simultaneity is relative

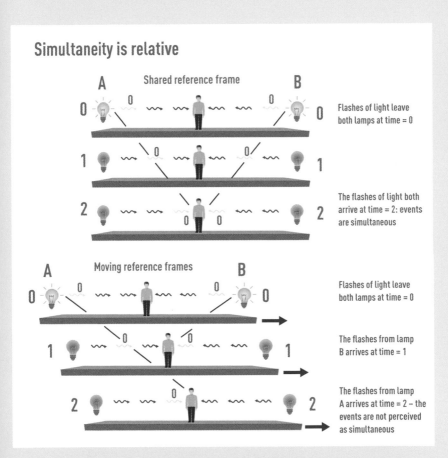

Shared reference frame

A   B

Flashes of light leave both lamps at time = 0

The flashes of light both arrive at time = 2: events are simultaneous

Moving reference frames

A   B

Flashes of light leave both lamps at time = 0

The flashes from lamp B arrives at time = 1

The flashes from lamp A arrives at time = 2 – the events are not perceived as simultaneous

This simple thought experiment shows how people have different experiences of events. Here, the location of the light flash at any given time is marked '0', demonstrating that our perception of how time is flowing for different objects depends on their relative motion.

10.3 Time dilation

Special relativity suggests that time flows differently for objects in relative motion.

Special relativity showed how relative motion affects time. The effect is only obvious at speeds close to the speed of light, but precision measurements have shown that it really happens.

To think about time, Einstein imagined a clock made of light. A pulse of light would act like a pendulum, bouncing between two mirrors. Each trip up and down would measure one tick.

If the clock was in motion and you were standing still and observing it, you would see the pulses of light trace out triangles as it moved along. The light would then have to travel further, so it would take longer to move between the mirrors. From your perspective, the clock would tick more slowly.

This concept of **time dilation** has been confirmed in experiments. In 1971, a highly sensitive atomic clock was flown around the world in a commercial aeroplane, while an identical clock remained on the ground. The difference between them corresponded to the amount expected due to special relativity and was tiny – around 100 billionths of a second.

The effects of time dilation due to motion, particularly on satellites, are complicated by the way that gravitational fields also distort time (see page 242).

Time dilation has inspired many thought experiments such as the twin paradox. Here, one twin stays on Earth while the other ventures into space and then returns (see opposite). When the space explorer returns, she is younger than her twin, having aged less owing to her journey.

Time-travelling twins?

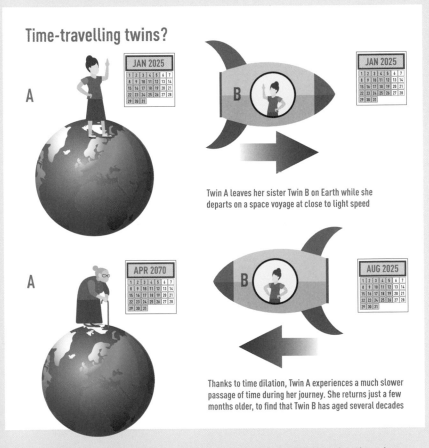

Twin A leaves her sister Twin B on Earth while she departs on a space voyage at close to light speed

Thanks to time dilation, Twin A experiences a much slower passage of time during her journey. She returns just a few months older, to find that Twin B has aged several decades

The central question of the twin paradox is why one sister ends up older than the other: since motion is relative, shouldn't Twin A also see Twin B ageing more slowly on Earth? The solution lies in the fact that Twin A changes reference frames between outward and homeward legs of her trip.

10.4 Length contraction

An object approaching the speed of light appears shortened in the direction of travel.

Time dilation (see Topic 10.3), has a counterpart in space: **length contraction**. Fast-moving objects become compressed in the direction of their motion. Again, the difference is tiny for normal speeds.

For example, imagine measuring a spacecraft's length by firing a laser from one end to the other. The laser will have a constant speed (c), and so the astronauts on board can easily calculate the length of their ship. As seen by a distant observer on Earth, however, time dilation causes time on the ship to slow down. If the laser spends less time in transit between one end of the spacecraft and the other, but still travels at the same speed, it isn't going as far. The measured length of the spacecraft will be shorter.

As with all relativistic effects, the contraction only becomes significant near the speed of light. Even at 44,000 km/s (100 million miles per hour), an object will still have 99 per cent of its proper length.

The phenomenon has not been measured directly, but it does explain the results of the Michelson–Morley experiment (see Topic 10.1). Earth's motion through space caused their interferometer (Topic 2.7) to contract slightly in one direction, neatly cancelling out the expected difference in travel times for its two light beams.

Both time dilation and length contraction are governed by an equation called the Lorentz factor.

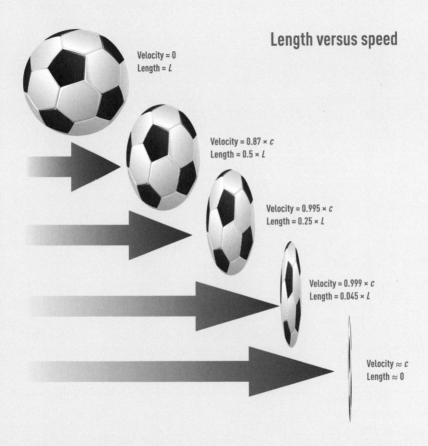

Length versus speed

Velocity = 0
Length = L

Velocity = $0.87 \times c$
Length = $0.5 \times L$

Velocity = $0.995 \times c$
Length = $0.25 \times L$

Velocity = $0.999 \times c$
Length = $0.045 \times L$

Velocity $\approx c$
Length ≈ 0

The phenomenon of length contraction only becomes noticeable at
extreme relative velocities, as shown in this comparison of a soccer
ball kicked at different speeds.

10.5 E=mc²

Objects accelerating close to the speed of light gain mass rather than speed.

The most famous equation in all of physics equates energy and mass. It was yet another consequence of the fact that all observers measure the same speed of light, and it is the mechanism behind the creation of the universe.

If the speed of light is the ultimate speed limit, how can it be reconciled with the conservation of momentum? For example, if energy is supplied to a moving object going almost as fast as the speed of light, its momentum should increase (see Topic 1.3). Einstein's leap was to say that, because the object can't go faster than light, its mass increases instead. It would take infinite energy to accelerate it to the speed of light, which is impossible.

Mass and energy, which seem like completely different things, are different aspects of the same phenomenon. Einstein was able to use these relationships to derive the famous equation showing precisely how they can be interchanged. Energy equals mass times the speed of light squared: $E=mc^2$.

When discussing fast-moving particles, physicists often distinguish between an object's rest mass and its relativistic mass.

When atoms split and the resulting debris has less mass than the original, the difference in mass is released as energy. This is exploited in nuclear energy and atomic bombs. The reverse is true in particle accelerators (see Topic 9.3), which create new particles out of energy. This process also happened at the very beginning of the universe, when mass was created from energy in the big bang.

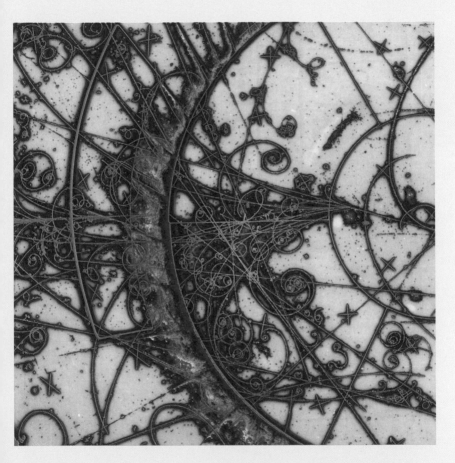

Particle accelerators rely on the equivalence of mass and energy – colliding heavy subatomic particles at high speeds liberates large amounts of energy, which then condense back into a variety of different exotic particles leaving tracks like the ones shown above.

10.6 Spacetime

The three separate
dimensions of space
and one of time combine
to make a flexible four-
dimensional manifold.

Special relativity (see Topic 10.2) shows that space and
time are linked. Further development of this idea explains
how gravity works and gives us one of the world's most
useful gadgets.

Newtonian physics saw space as a cube-like map with three
dimensions. But in the aftermath of Einstein's 1905 publication
of special relativity, his former tutor Hermann Minkowski
developed a new way of looking at the universe: **spacetime**.

In Minkowski's model, three space-like dimensions and
one time-like dimension are intimately connected in a four-
dimensional structure. The effects of relativistic motion can
be treated as creating a rotation in the coordinate system of
spacetime, such that the length dimension in the direction of
travel is shortened while time is extended.

It's impossible to imagine such a four-dimensional structure,
but you can envisage it as a two-dimensional sheet shown
opposite. Einstein took the idea of spacetime forwards to show
how, when it is warped, it gives rise to the force of gravity.

In orbit above Earth, the curvature of spacetime due to Earth's
mass is less than it is on the surface, and so clocks tick faster
than they do on the ground. This effect has to be taken into
account in GPS satellites; otherwise, they would not be able to
give us an accurate location.

Einstein initially thought
that spacetime was just a
neat mathematical trick,
before realizing that it
described reality.

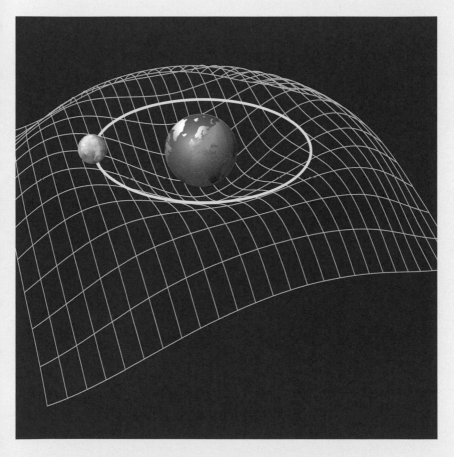

Rubber-sheet visualizations of spacetime often depict the universe as a flat, flexible sheet distorted (according to general relativity) by the presence of mass. This example shows how the Moon's orbit is defined by the gravitational well around Earth.

10.7 General relativity

Gravity creates
conditions equivalent
to constant acceleration,
and shows that mass
distorts spacetime.

Isaac Newton wrote that he would 'leave to the consideration of the reader' the explanation of how his law of universal gravitation worked. Albert Einstein provided the answer.

The starting point for Einstein was to realize that, in certain situations, you can't tell whether you are experiencing uniform acceleration or the force of gravity. An example is a person in a rocket accelerating in space at exactly 9.81 m/s/s (32.2 feet per second per second). If a man inside drops a ball, it would fall in just the same way as if he was standing on Earth. This is the **equivalence principle**.

Einstein realized this property of gravity could be replicated using the mathematics of curved surfaces. Empty spacetime would be flat, but objects with mass and energy could cause it to warp (see Topic 10.6). The resulting general theory of relativity, published in 1915, therefore showed how gravity works. Just as there are no straight lines on the surface of a sphere, objects like planets around the Sun still follow the shortest path through spacetime – that path just happens to be curved.

General relativity
predicts the existence
of gravitational waves
created by sudden
changes in the mass
of objects or systems.

He demonstrated the power of his theory by using it to solve the long-standing problem of Mercury's orbital precession (see Topic 10.1), but it was only with the confirmation of another consequence – gravitational lensing (Topic 10.8) – that his theory became widely accepted.

The equivalence principle

Inside a sealed box, physical laws are the same whether the box is in a gravitational field, or subject to constant acceleration ...

Apple falls due to gravity

Apple falls due to acceleration

... so if gravitational fields and acceleration are the same, shouldn't extreme gravitational fields bend the path of light in the same way as extreme acceleration?

Light beam deflects due to acceleration

Light beam bends due to gravity

One simple implication of general relativity is that gravitational fields should bend light in the same way that acceleration does. In normal circumstances, both effects are unnoticeable, but this bending of light has now been observed in a variety of situations.

10.8 Gravitational lensing

Large masses such as stars bend the paths of light rays passing nearby.

A few months after the end of World War I, British astronomer Arthur Stanley Eddington photographed stars during an eclipse of the Sun. The results confirmed one of the predictions of general relativity – and a phenomenon that would prove hugely useful for future astronomers.

Eddington had travelled to the African island of Principe, one of the locations from which the eclipse of 29 May 1919 could be observed. The eclipse gave Eddington the chance to photograph stars close to the Sun, to test whether or not large masses could bend rays of light.

The eclipse allowed Eddington to see stars that would not normally be visible in daylight. Comparing photos taken during the eclipse with the same patch of sky at night, he could see that some stars had indeed shifted position.

Why should this happen when photons of light have no mass? The reason is because large masses like the Sun create dents, or gravitational wells, in spacetime (see Topic 10.6). Light's path is deflected by the wells, like a soccer ball on bumpy ground.

The most distant galaxy so far identified using gravitational lensing lies some 13.2 billion light years from Earth.

Today, the deflection of light by gravity, called **gravitational lensing**, is used to detect distant objects whose light is bent by nearer ones. Lensing has a magnifying effect, revealing galaxies that would otherwise be too faint to see.

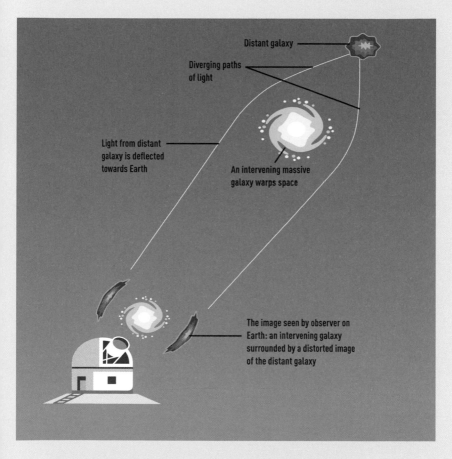

Distant galaxy

Diverging paths
of light

Light from distant
galaxy is deflected
towards Earth

An intervening massive
galaxy warps space

The image seen by observer on
Earth: an intervening galaxy
surrounded by a distorted image
of the distant galaxy

Gravitational lensing magnifies and intensifies light from very distant galaxies, but it
can also reveal the mass of the object responsible for the lensing. This helps confirm
that about 85 per cent of the mass in our universe is mysterious and invisible dark matter.

10.9 Singularities and black holes

Superdense points exist in space where the laws of physics no longer make sense.

The idea that an object could be so massive that not even light could escape its gravity was proposed in the 18th century. It was only after the publication of general relativity that the idea was taken seriously and the hunt to find an example of the phenomenon began.

In 1916, Karl Schwarzschild published an analysis of Einstein's model of space, time and gravity, showing the possibility of mass concentrated in a single point. In this situation the normal rules of spacetime break down in a so-called **singularity**.

In 1931, Subrahmanyan Chandrasekhar proposed that singularities might form through the collapse of a star. Further studies showed that a singularity would seal itself off, creating an **event horizon** barrier around it at the point where gravity became strong enough to overcome the speed of light. The term **black hole** was coined in the 1960s.

A relative of the black hole is the **wormhole**, or Einstein-Rosen bridge, a shortcut between two areas of spacetime. It's created where a black hole's gravitational well punches through to another location, forming a **white hole**.

White holes remain elusive, but many black hole contenders have been found – detected indirectly from their gravitational tug on other stars, their emission of powerful jets of matter and the fact that they glow in X-ray radiation.

The closest known black hole, called Cygnus X-1, lies about 6,070 light years from Earth.

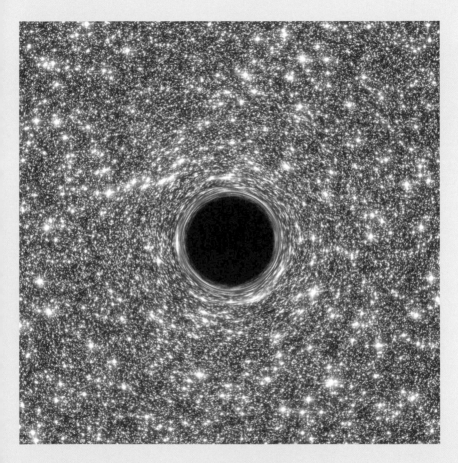

This artist's simulation shows a supermassive black hole in the crowded central regions of a galaxy like our own. Such monster black holes, with a mass equivalent to millions of Suns, are thought to form the seeds around which galaxies form.

10.10 Time machines

Does Einstein's theory of general relativity mean that we can travel through time?

One of the most intriguing questions to arise from general relativity is whether the flexibility of spacetime makes time travel possible. Could we really visit the future or go back in time to see how our ancestors lived?

Some solutions to Einstein's equations do permit the creation of 'closed time-like curves' connecting different parts of the universe. Since the fabric of spacetime binds space and time together, you'd be visiting another location in time as well as another location in space.

Wormholes (see Topic 10.9) might act like this, permitting time travel if the singularities at their heart could be avoided. It might even be possible to harness wormholes in the universe. By anchoring one mouth and dragging the other across space at relativistic speeds, you could create a time differential between the two ends.

Cosmologist Stephen Hawking and others believe time travel must be forbidden by physics to preserve sense and order in the universe. On 28 June 2009, Hawking held a party for time travellers. But he only publicized the party after the event, as then the only guests would be people who had travelled back in time to visit him. No one came and Hawking was left to party on his own.

Cosmologist Kip Thorne came up with the physics of the wormhole to help his friend Carl Sagan write a science-fiction novel.

Travelling through time

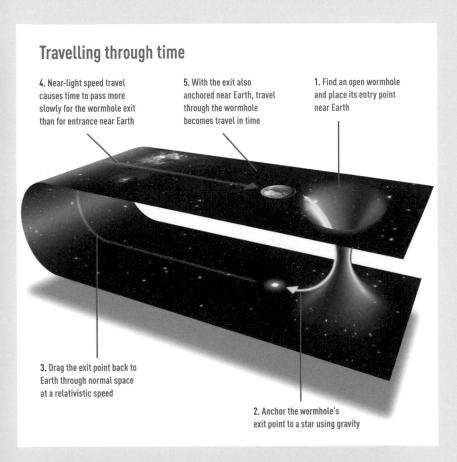

4. Near-light speed travel causes time to pass more slowly for the wormhole exit than for entrance near Earth

5. With the exit also anchored near Earth, travel through the wormhole becomes travel in time

1. Find an open wormhole and place its entry point near Earth

3. Drag the exit point back to Earth through normal space at a relativistic speed

2. Anchor the wormhole's exit point to a star using gravity

One way of building a time machine is to take advantage of an open wormhole between our region of space and another. If wormholes exist and can be travelled through, then the rest is simply a technological challenge – though one far beyond humanity's current capability.

Glossary

Boson
Any subatomic particle that is immune to the Pauli exclusion principle. Bosons include force-carrying particles such as photons.

Current
A flow of charge-carrying particles through a conducting material. Charge-carriers are usually negatively charged electrons, but current is, by convention, treated as if it was carried by positive carriers.

Electromagnetic radiation
A moving disturbance transferring energy across space and consisting of perpendicular electrical and magnetic waves that reinforce each other.

Electron
A widespread, negatively charged and low-mass lepton particle found orbiting the positively charged nucleus of an atom. The transferring or sharing of electrons is responsible for bonding atoms together, and freely moving electrons are the carriers of electric current.

Fermion
Any subatomic particle that obeys the Pauli exclusion principle. Fermions include all common matter particles.

Frame of reference
Any fixed system of coordinates for measuring physical events. Einstein's theories of relativity explain how measured properties vary when two frames of reference are in relative motion or affected by gravitational fields.

Fundamental force
Any of four forces responsible for all known interactions in physics. They are gravitation, electromagnetism, the strong nuclear forces and the weak nuclear forces.

Gravitation

An attractive force created by objects with mass, which causes other objects to accelerate towards them. According to general relativity, gravitation arises from a distortion of space and time around massive objects.

Inertia

An object's innate tendency to resist change to its motion when subjected to a force, linked to its mass.

Ionising radiation

Particles or high-energy electromagnetic radiation released during the decay of unstable radioisotopes into more stable forms. Ionising radiation may be either heavy alpha particles, lightweight beta particles or massless gamma rays.

Lepton

Alongside quarks, one of the two types of fundamental particle that make up all matter. Leptons include three generations of negatively charged particles (the electron, muon and tau), and their accompanying low-mass, neutral particles called neutrinos.

Mass

A measure of the amount of matter contained within an object, linked to its inertia. Mass is thought to be created by matter interacting with the Higgs field.

Momentum

A property calculated by multiplying an object's mass by its velocity. Momentum is conserved during collisions.

Neutron

An uncharged subatomic particle found alongside protons in the atomic nucleus and contributing substantially to its mass. Each neutron in turn consists of three quark particles.

Nuclear fusion

The process of joining together nuclei of light elements such as hydrogen to create heavier nuclei and release energy, similar to that which takes place in the core of the Sun.

Nuclear fission

The process of splitting apart the nuclei of heavy atoms to produce lighter ones, usually releasing energy as a by-product.

Orbital shell

A specific region around an atomic nucleus in which electrons can be found. The distribution of electrons throughout various orbital shells explains a wide range of atomic properties including the chemical reactions of different elements.

Pauli exclusion principle

A physical law operating on a very small (quantum) scale that prevents two fermion particles in a system from taking on exactly the same properties.

Photon

A bullet-like packet of electromagnetic waves that allows light to pass through a vacuum. Photons also act as bosons in the transfer of electromagnetic energy.

Potential difference

A difference in electrical potential energy between two points in a circuit, measured in volts.

Proton

A positively charged subatomic particle found in the atomic nucleus. Protons are made of three quarks and have a similar mass to neutrons.

Quark

One of two types of fundamental particle that make up all matter. There are six quarks in total but only two, the up and down quarks, are found in everyday atoms, as the components of protons and neutrons.

Radioisotope

Any form or isotope of an atom in which an excess of neutrons in the nucleus creates instability. Radioisotopes decay towards more stable forms through the release of ionising radiation on a variety of timescales.

Uncertainty principle

A physical law operating on the quantum scale that prevents certain pairs of conjugate properties (such as position and momentum) being measured simultaneously with absolute precision.

Virtual particle

Any particle that can be briefly created out of nothing thanks to the uncertainty principle linking time and energy. The exchange of virtual boson particles transfers fundamental forces between subatomic particles.

Index

Acknowledgements

Author's dedication

Graham Southorn: For Ruth Southorn.

Picture Credits

Quantum Books Limited would like to thank the following for supplying the images for inclusion in this book:

7: CERN; 19: Shutterstock/conrado; 33: DR GARY SETTLES/SCIENCE PHOTO LIBRARY; 53: SCIENCE PHOTO LIBRARY; 63: ESA/NASA/SOHO; 71: Shutterstock/Vit Kovalcik; 77: SHEILA TERRY/SCIENCE PHOTO LIBRARY; 81: VICTOR DE SCHWANBERG/SCIENCE PHOTO LIBRARY; 89: Shutterstock/Vlue; 91: NASA Earth Observatory image by Jesse Allen and Robert Simmon, using EO-1 ALI data from the NASA EO-1 team; 95: NASA images courtesy Goddard Space Flight Center Ozone Processing Team.; 97: Shutterstock/Puwadol Jaturawutthichai; 105: Argonne National Laboratory; 119: Shutterstock/WHITE RABBIT83; 125: Shutterstock/Albert Russ; 127: Shutterstock/Papa Bravo; 135: Shutterstock/Miroslav Fechtner; 149: VOLKER STEGER/SCIENCE PHOTO LIBRARY; 153: Dave Jones; 171: Photo courtesy of National Nuclear Security Administration / Nevada Site Office; 177: LAWRENCE BERKELEY LABORATORY/SCIENCE PHOTO LIBRARY; 183: Shutterstock/lafayette-picture; 197: Shutterstock/SpeedKingz; 201: Shutterstock/Naeblys; 211: CERN; 225: CERN; 239: CERN; 247: NASA, ESA, D. Coe, G. Bacon (STScI); 249: DETLEV VAN RAVENSWAAY/SCIENCE PHOTO LIBRARY.

While every effort has been made to credit contributors, Quantum Books Ltd. would like to apologize should there have been any omissions or errors and would be pleased to make the appropriate corrections to future editions of the book.